TIDAL WATER

A History of Wellfleet's Herring River

By

John W. Portnoy

Alice M. Iacuessa

Barbara A. Brennessel

With a foreword by John T. Cumbler

Published by

Friends of Herring River

Wellfleet, Massachusetts

Published by Friends of Herring River
Wellfleet, Massachusetts
www.friendsofherringriver.org

Copyright © 2016 Friends of Herring River
Printed in the United States of America

ISBN-13: 978-1534829022
ISBN-10: 1534829024

All rights reserved, no parts of this book may be reproduced in any form by any means, graphic, electronic, or mechanical, including photocopying, recording, taping, or by any information storage retrieval systems without the written permission of the authors or publisher except by a reviewer who may quote brief passages in a review. Members of educational institutions and organizations wishing to photocopy any of the work for classroom use, or authors and publishers who would like to obtain permission for any of the material in the work, should contact Friends of Herring River or the authors.

Copies of this book may be purchased at select on-line sellers or select bookstores.

First Edition
July 2016

Table of Contents

Foreword — v

Acknowledgements — vii

Chapter 1. Post-glacial evolution of the Herring River valley, 18,000 years ago to 1600 AD — 1

Chapter 2. Native American settlement and land use, about 10,000 years ago to late 1700s — 9

Chapter 3. Early white settlement of the Billingsgate Islands: farming, fishing, saltworks, 1640-1800 — 21

Chapter 4. Resource depletion, 1600s-1800 — 39

Chapter 5. Mosquitoes and the diking of the River, late 1800s-1909 — 43

Chapter 6. The diked Herring River, 1909-1960s — 53

Chapter 7. Shift to tourism and summer residents, late 1800s-1950s — 61

Chapter 8. Dike failure, controversial reconstruction and ecological assessments, late 1960s through 2000 — 73

Chapter 9. Planning for tidal restoration, 2004 to the present — 87

Chapter 10. Conclusion — 91

Literature Cited — 93

About the Authors — 103

UNLESS someone like you

cares a whole awful lot,

nothing is going to get better.

It's not.

Dr. Seuss, *The Lorax*

Foreword

A century and a half ago Henry David Thoreau prophesized that the Herring Rivers on Cape Cod would be "more numerous than herrings." Just five decades later the Town of Wellfleet took a vote that helped Thoreau's ironic prophecy come closer to reality than even he had imagined. With that vote in 1908 the people of Wellfleet, against the objections of a minority of its citizens, set in motion the diking off of the Herring River from salt water tidal flushing. Behind the dike across the river mouth, saltwater marshes, which had sustained farmers for generations, died. Oyster beds disappeared and very few herring made it upstream to spawn in the freshwater ponds at the head of the river.

The majority of the people of Wellfleet voted to build the dike because the town's leading figures, particularly Lorenzo Dow Baker, argued that mosquitoes and marshland were holding the town back. Baker claimed that the future for the town lay in tourism and tourists would not come to Wellfleet if the area were plagued with mosquitoes. Mosquitoes were very much in the news in the first decade of the new century. In 1902 Walter Reed led the team that established that yellow fever was caused by mosquitoes and in 1905 the US sanitary officers put in place a program of draining marshes and eliminating standing water around the Panama Canal project, which dramatically reduced the death rate among the workers on the project. Concerned that mosquitoes were a danger to health and could be dramatically reduced by draining marshes, Wellfleet, after resisting for a few years, finally agreed to dike the river to reduce mosquitoes and open up more land for agriculture and tourist development.

Lorenzo Dow Baker and others in town who believed that the future lay in tourism were unaware of the unintended consequences of diking the river. Oyster beds upriver from the dike died. Fishermen who kept their boats above the dike were forced to move. The rich salt hay marshes, which supplied winter feed for farmers' cattle, were gradually overrun by a dense shrub thicket. And contrary to the predictions of the dike supporters, the mosquitoes did not go away. Indeed, without the flush of the tides bringing both oxygen-rich sea water and small fish to feed upon the mosquito larvae, the problems of mosquitoes only increased. Upriver from the dike the land dried, freshwater plants moved into the wetlands and dry land plants took over large swaths of land. Without the rebuilding of the marsh through tidal action, the land subsided and acidified. And of course the annual run of herring which had sustained a significant town fishery all but ended.

Henry David Thoreau meant his comment about more Herring Rivers than herring to be facetious, but the town of Wellfleet, at the urging of some of its most important citizens, did its part to move his prophecy closer to reality. Today the people of Wellfleet are in a position to change that and restore the river to its more natural role.

John T. Cumbler
Professor Emeritus
University of Louisville

Wellfleet, MA
July 2016

Acknowledgements

Don Palladino, President of Friends of Herring River (FHR) planted the seed for a book about the history of the Herring River. He realized that some of the citizens of Wellfleet and Truro are well aware of the project to restore tidal flow to the River, while others are still learning about the proposed restoration project. Don thought that it was important to reveal the early history of the region and the reasoning behind the decisions, made in the early 1900s, which led to the building of the Chequesset Neck Dike.

FHR Board member Lisbeth Chapman, environmental historian and author John Cumbler, and National Park Service Historian Bill Burke were kind enough to read the manuscript and provide valuable suggestions. The authors appreciate access to the archives of the Wellfleet Historical Society and Museum as well as the Cape Cod National Seashore and the resulting treasure trove of valuable information and photographs, which helped to tell this story. We also are grateful for permission to use images from both sources; photographs are from the Historical Society unless otherwise noted.

The Wellfleet Cultural Council provided funding to get the book off the ground. Bill Iacuessa assisted with technological expertise. Marisa Picariello designed the cover. We offer a special thanks to Susan Dunn for her Herring River Restoration poem.

Chapter 1. Post-glacial evolution of the Herring River valley, 18,000 years ago to 1600 AD

Glacial origins

The history of the Herring River valley really begins about 18,000 years ago when the Laurentide Ice Sheet, the last glacier to cover this region, began to "retreat," actually to melt, from south to north. At this time so much of the Earth's surface water was locked up in glaciers that sea level was about 300 feet lower than today, while the ice cover of Cape Cod was more than 1500 feet thick.

The outer Cape formed from sediment deposited directly from a huge lobe of the melting glacier to the north and east (the South Channel Lobe), and from sediment carried in meltwater streams running off the retreating glacier, and flowing generally westward, to form the Wellfleet outwash plain. In addition, thick layers of clay settled into the bottoms of large lakes that were, for a time, impounded between the outwash deposits and the retreating glacial front. Evidence of these glacial lakes is present in the Herring River area, e.g. the exposed clay sea cliff just south of Newcomb Hollow Beach, and in deep underground clay deposits near Gull Pond.

Outwash plains, as one would expect given their deposition by meltwater streams, are generally fairly flat; however, the land surface of north Wellfleet and around the Herring River valley is hilly with deep "kettle" holes. This "knob and kettle" or "kame and kettle" topography resulted from the accumulation of sediment around melting ice, i.e., the melting glacier. Thus, the hills forming Great, Griffin, Merrick, and Bound Brook Islands are kames, a land form produced when sediment flowing in streams across the glacier surface collected in holes in the ice; when the ice completely melted away, the collected sediment stood above the rest of the outwash plain as hills. Kettle depressions, including those presently containing Gull, Higgins,

Williams, and Herring Ponds, formed through the inverse process: remnant blocks of ice left on the outwash plain after retreat of the main ice front became buried in outwashed sediment. Eventual melting of these ice blocks left depressions in the outwash deposits.

The Herring River valley is too broad to have been cut by the modern Herring River. It formed more likely through "spring sapping" or the seepage of groundwater through the meltwater-deposited sediment (Oldale 1992). This could have occurred during the stage of glacial retreat when a lake of meltwater was impounded between a glacial ice front to the east (the South Channel lobe), and the already-deposited mound of outwash sediment forming the outer Cape to the west. If this lake level were higher than the land surface on the opposite side of this mound, then groundwater would flow westward through the mound, essentially a leaky dam, and spill out through the land surface as springs. The springs themselves would tend to migrate uphill as they eroded the outwash from which they flowed; meanwhile their water would flow downslope on the westward tilting outwash plain, eroding roughly east-west oriented valleys toward modern Cape Cod Bay. It is conjectured that this process produced the Herring River valley and the other so-called "pamets," a geologic term named after Truro's Pamet River. A modern equivalent of this process, at a much smaller scale, can be seen on beaches at low tide where groundwater breaks through the sloping beach face (Fig. 1), eroding miniature valleys similar to the pamets.

During this period of glacial retreat and pamet formation, sea level was about 300 feet lower than today, and what was to become the Herring River valley was high and dry. It would not experience seawater flooding for thousands of years, that is, when post-glacial sea-level rise finally reached the Herring River valley. The first (i.e., deepest) wetland sediments in the Herring River flood plain formed about 3000 years ago (Roman 1987).

Figure 1. Groundwater discharging onto the beach face at low tide simulates in miniature the probable creation of the Herring River valley by spring sapping through glacial outwash (see text) (photo by John Portnoy.

Early vegetation of the Herring River watershed

Radio-carbon dating of the oldest organic matter found in the bottom of Wellfleet's kettle ponds indicates that the bleak landscape remained cold, dry, wind-swept, and, consequently, devoid of vegetation for several thousand years after the glacier melted away (Winkler 1985; Oldale 1992). For much of this time, the towering glacial ice front loomed to the north, exerting a strong influence on the Cape's climate (Winkler 1985).

Extensive coring and pollen grain analysis of the sediments of the kettle ponds (especially Duck and Dyer Ponds in Wellfleet, and Great

Pond in Truro) by Marjorie Winkler of the University of Wisconsin showed the evolution of upland vegetation surrounding the Herring River valley since the glaciers retreated. Sometime before 12,000 years ago, ground-hugging mosses, arctic willow (*Salix* spp.) and spruce (*Picea* spp.) began to colonize a cold tundra landscape, with cattail (*Typha* spp.) and sedges (e.g., *Scirpus* spp.) in wet areas (Winkler 1985). With a warming climate between 12,000 and 10,500 years ago, the low spruce parkland was replaced by a boreal forest of spruce, jack pine (*Pinus banksiana*), and green alder (*Alnus viridis*). Between 10,500 and 9000 years ago, fires became frequent and the forest shifted to jack pine and white pine (*Pinus strobus*), with extensive heathlands (Ericaceae) probably along exposed sea cliffs. Since then the Cape forests probably have looked much like they do today, dominated by pitch pine (*Pinus rigida*) and oaks (*Quercus* spp.), with bayberry (*Myrica pensylvanica*), winterberry (*Ilex verticillata*) and roses (*Rosa* spp.) replacing the rapidly disappearing heathland plants in more open areas. Maple (*Acer* spp.), beech (*Fagus grandifolia*), and other hardwoods grew in the moister kettle holes. Hemlock (*Tsuga canadensis*) and beech were apparently quite abundant until about 4700 years ago, when the former disappeared and beech declined.

The appearance and likely influence of Native American settlement of the region is evident in pond sediments containing corn (*Zea mays*) and sorrel (*Rumex)* pollen deposited at least 200 years before European settlement. Changes in vegetation after the arrival of the Pilgrims in the mid-1600s are clear in the sediment record, and include increases in pollen of herbaceous species like grasses, goldenrods (*Solidago* spp.), and cultivated grains as forests were cleared for agriculture and building construction (Winkler 1985).

Development of Herring River wetland vegetation

Analysis of root and rhizome remains in deep sediment cores shows that salt marsh habitats, including tidal flats and both low and high salt marsh (dominated by the salt-tolerant grasses *Spartina alterniflora* and *Spartina patens*, respectively), were found in all areas of the estuary including east of the present Route 6 and approaching Herring Pond (Fig. 2) (Roman 1987; Winkler 1994). In the upper reaches of the estuary, these habitats went through a series of changes over the past 200 years: from tidal flat to low salt marsh, then high salt marsh, forbs (broad-leafed herbaceous plants) and cattail, cattail and shrubs, and finally shrubs and grasses as sediment gradually filled in the Herring River valley (Fig. 3). Also, inlets to Cape Cod Bay at Ryder Beach (Bound Brook) and Duck Harbor were probably beginning to close before European settlement in the 1640s, reducing the inflow of sea water. Inlet closure started as a natural process caused by reduced tidal volume with the infilling of bays and by increased sediment deposition and the extension of the barrier beaches along the Cape Cod Bay shore; the two processes reinforced each other in a positive feedback. However, deforestation, intensive grazing by domestic animals, and consequent upland soil erosion after European settlement in the 18th and 19th centuries may have accelerated the shoaling of inlets and bays at Bound Brook and Duck Harbor (See Chapter 5).

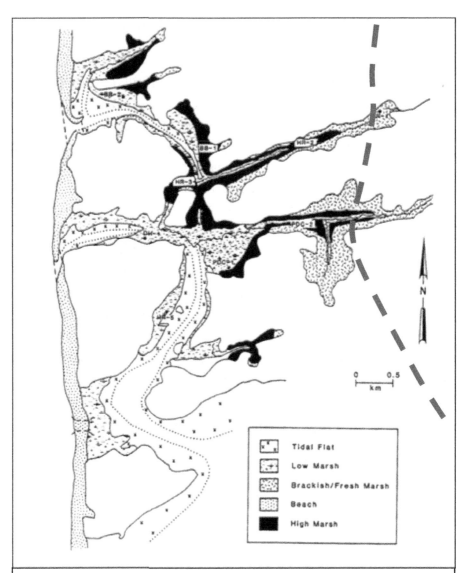

Figure 2. Map of pre-historic wetland and beach habitats reconstructed from the interpretation of sediment cores (Roman 1987). Inlets at both Ryder Beach and Duck Harbor were still open to tidal exchange during the period depicted, probably about 500 years ago. Salt marsh habitats extended throughout the system, even east of the present Route 6 (dashed line); there were much more open-water and intertidal areas than today.

Figure 3. Looking southeast from Bound Brook Island across the Herring River salt marshes about 1903, shortly before tides were blocked at the river mouth. Merrick Island is in the background. Salt-marsh grasses covered the flood plain. Structures crossing the waterways were to block migrating herring so that they could be netted (See Chapter 3). The road paralleling the river meander follows the same track as Bound Brook Island Road today; however, river flow through this meander has been cut off since channelization in the 1920s and 1930s.

Chapter 2. Native American settlement and land use, about 10,000 years ago to late 1700s

Little is known about the Paleoindians who were the first humans to arrive on Cape Cod during the Early Archaic Period about 10,000 to 8000 years ago (Johnson 1997; Cumbler 2014). Although there are few artifacts from this time period, it is believed that Native American living on Cape Cod existed in small units as hunter-gatherers. These people were nomadic and lived in a land that only a few thousand years ago had seen the final withdrawal of glaciers. Their diet consisted of roots, nuts, and berries which they gathered, and game such as squirrel, beaver, and deer which they hunted. They also without doubt used the rich resources from the sea and inland ponds, rivers, and marshes. However, they lived in a changing environment. As the glaciers retreated, the climate changed; landforms were transformed with both the sea and land rising. New species arrived, perhaps even the extinct mastodon and animals now found much farther north such as caribou (Johnson 1997). Lack of artifacts and evidence of specific settlement sites from these very early periods are due to the fact that possible settlements would now be submerged because of rising sea levels and coastal erosion. For example, the Atlantic shoreline extended one hundred miles eastward to the area of what is now Georges Bank.

Late Archaic Period

Probably the first Native Americans to live in the area of the Herring River appeared about 5000 to 3000 years ago during the Late Archaic Period (Torp et al. 2013; Gillis and Herbster 2013), about the time of marsh establishment in the Herring River Valley (Roman 1987). This was a period of rapid sea level rise and the development of estuaries, places where freshwater rivers and streams meet saltwater bodies. Sea level rise and the development of estuaries led to an increase in oysters and crabs and large seasonal runs of anadromous fish. Estuaries on Cape Cod in general, and the Herring River estuary in particular, provided favorable locations for Native American settlements whether permanent or seasonal.

Numerous artifacts show that during the Late Archaic Period Native Americans followed a seasonal pattern of life (Johnson 1997). Their use of plant and animal resources depended on what was available and when it was available. During this time period one would have seen Native Americans seasonally exploiting the waters of the Herring River. The scene was of men and women collecting oysters (*Crassostrea virginica*) and scallops (*Argopectin irradiens*) and digging for clams (*Mercenaria mercenaria*). Since the Herring River estuary was also a prime fishing location for anadromous fish such as shad (*Alosa sapidissima*) and river herring, which include alewives (*Alosa pseudoharengus*) and blueback herring (*Alosa aestivalis*), there would have been fish weirs in the estuary (Fig. 4). These were made of netting, brush, or stakes anchored to the bottom and ending in a circular enclosure effectively trapping many fish at one time. At the head of the estuary with the rising tide, migrating fish came to feed, congregating as they paused to adjust to the change from saline to fresh water environments. As the tide receded the current pulled fish into the weir. The schooling behavior of river herring made them particularly easy to catch in a weir. Different fishing techniques were used depending on the catch. For larger fish such as flounder (various genera), skate (various genera), dog fish (*Squalus acanthius*), and cod (*Gadus morhua*), Native Americans used nets made from plant fiber, baited hooks, spears and arrows made from bone or sharpened stone (Massachusetts Historical Commission 2014) (Fig. 5). For these larger fish and, especially in the case of striped bass (*Morone saxitalis*), Native Americans took advantage of the fact that on a high tide these fish swim into saltwater creeks and, as the tide ebbs, head back to deeper water. The Native Americans either blocked the channel

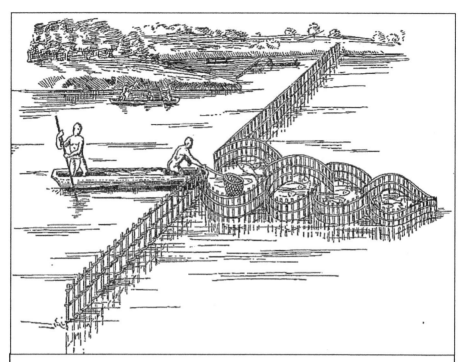

Figure 4. Native Americans practicing weir fishing. Weirs made of netting, brush, or stakes were anchored to the bottom. Weirs included a circular enclosure that trapped many fish at one time. (figure from www.americanindianshistory.blogspot.com)

funneling the fish into nets or speared them as they passed through a narrowed opening.

The Herring River valley would also have provided a productive environment for hunting. The woodlands, ponds, wetlands, and saltwater ecosystems supported a wide variety of waterfowl, shorebirds, and small mammals which were a food source for the Native Americans. The mammals also provided skins and the wild fowl small bones and feathers for ornamentation (Gillis and Herbster 2013; Johnson 1997; McManamon 1984). The native populations also gathered the abundant berries, roots, and nuts that grew in the area.

Figure 5. Native American fishing techniques. Native Americans used nets made from plant fiber, baited hooks, spears and arrows made from bone or sharpened stone to catch larger fish (figure from Braun and Braun 1994).

Late Woodland Period

During the Woodland Period about 2500 years ago, Native American settlements had increased, contained larger populations, and grew more sedentary. New aspects of life were introduced – funereal rites, agriculture, and fire. In 1979, at a construction site, an ossuary, a pit where Native Americans put the bones of their dead, was uncovered on Indian Neck just south of the Herring River. Ossuaries are normally associated with life in a permanent settlement. The one on Indian Neck contained many Native American remains. Native Americans during this time period buried or cremated their dead. Later they recovered the bones and held a second ceremony and burial where they honored their ancestors. Archeologists interpret this as evidence that these people lived in close-knit groups in the same area over many generations (Johnson 1997).

It is generally believed the New England native populations were nomadic during the Woodland Period, moving in groups to exploit seasonally available food sources. For native peoples on the outer Cape, this interpretation has been questioned. The resources needed to support life were not located at great distances, and it is now thought that from at least 2000 years ago the native people lived in year-round settlements (Echeverria 1993; McManamon 1984). They were living in complex villages similar to those Samuel de Champlain sketched when he visited the outer Cape 1000 years later in 1606. Based on Champlain's drawings, which recent archeological research suggests were accurate (Johnson 1997), the scene at the Herring River valley would have been one of a small village with domed dwellings surrounded by agricultural fields (Fig. 6). Native Americans had begun to select seeds for sowing, to weed their fields, and to use selective burning to encourage the growth of certain plants, all of which were key to an emerging native agriculture (Johnson 1997; Massachusetts Historical Commission 2014). These Native Americans would have grown crops such as the "three sisters"- corn, squash, and beans. These were grown in mounds fertilized with horseshoe crabs and fish such as alewives. Fire also was used to burn the underbrush in the forest to retain habitat for small mammals and birds and also to make hunting easier.

Nearby the settlements were mounds called middens. These were made up of discarded shells from mollusks that had been primarily a food source. In 2011 the National Park Service commissioned an archeological study of the newly acquired Baker-Biddle property on Bound Brook Island. This study uncovered more than 11,000 artifacts

Figure 6. Typical Native American Village as described by Champlain. Native Americans lived in small villages with domed dwellings surrounded by agricultural fields. They grew crops such as the "three sisters" – corn, squash, and beans that they fertilized with fish and horseshoe crabs.

such as pottery, bone, tool fragments, projectile points, and a large, mostly intact, midden. These remains are evidence of permanent Native American settlement on the site from the Late Archaic period through the Woodland Period (Torp et al. 2013). Another archeological study conducted by the National Park Service in 2012 identified middens in and near the Herring River valley. Two shell middens from this period were found on Great Island and other shell middens were found on Bound Brook and Griffin Islands and on the banks of the Herring River, one of which is on a positively identified settlement site (Gillis and Herbster 2013).

Supporting local folklore, there is also evidence in pond sediment that the native people created a channel (sluiceway) between Gull and Higgins Ponds (Winkler and Sanford 1994). The purpose for this was most likely to improve the herring run by allowing the fish to spawn in 90-acre Gull Pond, herring being an important source of food and fertilizer to Native Americans.

Contact Period

The Woodland Period lasted until 1500 AD when contact with Europeans brought changes to the Native American way of life. By 1500 European fishermen were beginning to fish the waters of the Grand Banks and to explore the coasts of the outer Cape. It is assumed that a number of indigenous people had contact with these fishermen. In 1602 the first recorded landing of a European on Cape Cod occurred when Bartholomew Gosnold disembarked at Provincetown. In 1605 and 1606 Samuel de Champlain explored Nauset Marsh and traded with the Native Americans. During his second exploration of the area Champlain sailed into Wellfleet Harbor. The presence of its extensive oysters beds caused him to name the area Port aux Huitres (Oyster Harbor) (Echeverria 1993; Torp et al. 2013).

The lifestyle of the Native Americans of the Herring River valley during the Contact Period remained very similar to that of the Woodland Period. They were organized in communal groups in well-defined areas (Gillis and Herbster 2013). Life continued to center near available natural resources needed for farming, fishing, gathering, and hunting. The Native Americans were loosely affiliated in a confederation as the Wampanoag and tended to live near major estuaries, with the Pamets in the Pamet estuary of Truro and the Nausets near Nauset Harbor in Orleans. The native people who lived in the Herring River estuary were known as the Punonakanits. It is difficult to know how many Native Americans lived in the Herring River estuary, but it has been estimated that the nearby Nauset population in Orleans and Eastham was at least 3700 to 4000 (Johnson 1997); the population in the Herring River valley was probably much smaller. Communities were linked by waterways and trails, some of which became present day roads such as Chequesset Neck Road (Johnson 1997) which most likely led to the Punonakanit villages in the Herring River valley.

Life for most Native Americans in the Northeast changed during the Contact Period. Devastating epidemics of diseases brought by Europeans occurred in New England from 1616 to 1619. However at this time Native Americans on the outer Cape lived some distance from European settlers, and the Punonakanits of the Herring River seemed to have escaped the worst of the epidemics (Torp et al. 2013). However, coinciding with the arrival of Europeans was a change in climate, the "Little Ice Age" (Johnson 1997). Winters became harsher and subsequently the growing season shorter. It is not clear how this change in climate affected the native populations but it is known that there were only 100 Native Americans in the area of modern Wellfleet by 1620 (Echeverria 1993; Lombardo 2000).

Plantation Period

By 1632 English settlement had spread from Plymouth to Cape Cod with the first permanent village at Sandwich. The first permanent settlement on the outer Cape started in 1644 with the purchase of land by a committee of men from Plymouth Colony. They bought land from local sachems extending from Pleasant Bay in Chatham to Truro, which they called the "Nauset Grant." Lands from Indian Brook, today Hatches Creek, to Bound Brook, which included the Herring River valley, were sold to the English by Lieutenant Anthony, sachem of the Punonakanits (Echeverria 1993). It appears that the native population did not have the same sense of land ownership as the English: when told by local Native Americans that the area of Billingsgate was not owned by any particular group, the Englishmen claimed it for their own. Native Americans viewed land as something which yielded necessary resources for the common good; the English saw land as wealth from property ownership (Pratt 1844). The English quickly moved into the southern part of the Nauset Grant, present day Orleans and Eastham. They cleared the land for agriculture and for grazing their livestock.

Despite expanding English settlement, the native people of the Herring River valley seemed to maintain their traditional way of life (Echeverria 1993). Although parcels of land on Bound Brook Island were allocated to English colonists as early as the 1660s and records show grants of twenty or more acres to Job Cole, John Smalley, Robert Wixam, Nicholas Snow, Richard Higgins, and Daniel Cole (Echeverria 1993), true European settlement in this area did not occur until the 1700s. In fact there were only a few colonial families between 1660 and 1675 (Gillis and Herbster 2013), and in 1674 it is estimated that the Punonakanits and neighboring Pamets had a population of seventy-two adults and a number of children (Echeverria 1993). The Punonakanits were also spared the ravages of King Philip's War of 1675-1678, which had severe consequences for other Native Americans in New England. The Punonakanits, like other native people of Cape Cod, adhered to an alliance with the Plymouth government and maintained neutrality during the war. Despite this neutrality, the off-Cape victory of the English ultimately gave them power to marginalize the native peoples of Cape Cod. As more English settled in the area the Native Americans were displaced (Massachusetts Historical Commission 2014).

The systematic practice of denying lands to the native people began as early as the 1670s. As the English community grew there was a competition for resources. Laws were passed that show a concern by the English settlers over the use of common land. In 1715 the town of Eastham, which included present day Wellfleet, started to allot land parcels and voted to grant certain parcels to the native people before allotting others to the English settlers. The lands allotted to the Native Americans were Indian Neck, Chipman's Cove, and land south of Fresh Brook. No lands were reserved in the Herring River valley for Native Americans. Although the town included the native population in the land distribution, the land was in fact parceled out in favor of the English settlers. The native people were deprived lands they needed to sustain their way of life. Because the native people lived a life following the cycles of nature, they required access to a number of seasonal resources. "The maintenance of this diversity was the very basis of their existence." (McManamon 1985, p. 47). The land allotted to them gave them access to rich shellfish beds, but they no longer had adequate access to coastal marshes, upland fields, and forests needed for farming and hunting.

Native Americans did a number of things to try to survive in the eighteenth century. They practiced small-scale agriculture and continued to hunt and fish. They lived on the edges of the English community and began selling their hand-crafted baskets, clothing, and brooms (Johnson 1997). One of the few economic opportunities available to the dwindling native people was employment by the English in the whaling industry but, in general, they fell into poverty. Diseases like smallpox also took their toll, especially in the smallpox epidemic of 1746-1748. By 1765 there were only sixteen Native

Americans living in Wellfleet (Johnson 1997). By 1792 that number had declined to three or four women and one man who lived in the Herring River valley. The last full-blooded Native American, Delilah Gibbs, died in 1838 (Echeverria 1993).

Chapter 3. Early white settlement of the Billingsgate Islands: farming, fishing, saltworks, 1640-1800

Early English agriculture in the Herring River valley

The Mayflower Pilgrims were predominantly farmers who in their new home planned to follow an agricultural lifestyle similar to that in England but "with much more godliness and greater dispersal of land ownership" (Cumbler 2014, p. 32). Although Provincetown, where they landed in November 1620, had excellent anchorage, and the waters teemed with finfish and shellfish, the Pilgrims originally did not intend to live lives based on the sea. Instead they were farmers and sought land, not sea resources. Thus, they decided not to settle on Cape Cod and chose Plymouth instead. Early on, however, they realized that Plymouth was far from an ideal place for a large settlement. The port at Plymouth was inadequate and the soil not rich enough to support a growing colony. New immigrants to the colony attracted by better soil had by 1639 established a village in Barnstable and by 1644 a committee was authorized to purchase land from the Native Americans on the outer Cape (See Chapter 2). These purchases ultimately included present day Eastham and Wellfleet, including the Herring River valley.

Land was parceled out to settlers as wood lots, pasture, farmland, and meadow grants. Eastham town records show that by 1696 a number of land grants mainly of upland and meadow were granted in the Herring River valley, specifically on Bound Brook and Griffin Islands. Since these grants were given to men such as John Doane and Daniel Cole who lived in Eastham, it is thought that these lands were not meant for settlement but for use as pastures or for the hay from salt meadows (McManamon 1985). The sale of hay from salt meadows was a lucrative industry; the hay was a source of food for livestock during the winter (Holmes 1995). From 1660 to 1703 there were six such meadows covering 22 acres around Bound Brook Island (Echeverria 1993) (Fig. 7).

Wellfleet, however, unlike Eastham and parts of Pamet, presently Truro, had poor soil for agriculture; therefore, settlement in Wellfleet was more likely determined by access to rivers and harbors than available land for farming (Rockmore 1979). Thus, among the first areas to be settled was the Herring River valley with its opportunities for navigation (Holmes 1995). Historians writing in the 19th century describe Bound Brook Island as being inhabited not by farmers but by whaling captains (Rockmore 1979).

The general assumption for Cape Cod is that as soil lost its fertility, settlers turned to maritime activities for their economy. For the Herring River valley, however, this may not be the correct assumption. Here with poor quality soil but good harbors, the 18th- and 19th-century residents depended on fishing as an economic base with only subsistence farming. The lack of plant agriculture is evident in the fact that in the 18th century, Truro grew more than enough grains and vegetables, while Wellfleet had to import most of the grain it needed to support the settlement (Rockmore 1979). (However, there is evidence of a tidal grist mill at Mill Creek (Massachusetts Historical Commission 1984)).

Thus the inhabitants of the Herring River valley, as the inhabitants of Wellfleet through much of its history, took advantage of any opportunity to earn a living. Evidence that people in the area engaged in a number of financial ventures can be seen in the probate records of Thomas Atwood of Bound Brook Island dated 1832. He owned farm animals, meadow lots and wood lots, part ownership in a dory boat and in a bass seine (Rockmore 1979). Further evidence is a property register of the Atwood-Higgins House in 1832, which lists both a dory

Figure 7. The extensive salt marshes and abundant salt hay for animal fodder and bedding were a major inducement for the Pilgrim's migration from their original settlement in Plymouth to Cape Cod (Cumbler 2014). The crop required no planting, cultivation, fertilization, or tillage. Salt hay harvest continued into the 1930s, about the time of this photo. Even today salt hay wrack is collected for garden mulch.

and livestock as part of its economic resources (Donaldson et al. 2010).

Even though upland farming was not successful in Wellfleet, one crop, cranberries (*Vaccinium macrocarpum*), was particularly well-suited for culture in freshwater wetlands. In 1816 a method of using sand to convert swampland into cranberry bogs was developed in Dennis, and a new commercial enterprise began. Although it was expensive to create these bogs, the financial return was excellent. By 1865

Wellfleet had 22 acres producing 55 bushels of cranberries (McManamon 1985). In 1950, a resident of Bound Brook Island described an old, abandoned cranberry bog on the south side of the road leading to the beach (Donaldson et al. 2010). Thus it can be assumed that cranberry growing as a commercial venture took place in the Herring River Valley in the 19th century. This was work that involved the entire family including children. Cranberries were harvested in September and October and children were frequently absent from school in order to work the bogs. Thus, in places such as Truro, the school calendar was adjusted to start after the harvest (Whalen 2007). One can assume the same for the school on Bound Brook Island.

Island School –Bound Brook Island

Before the Nauset Grant was purchased, Massachusetts Bay Colony passed the Education Laws of 1642 and 1647 requiring that children be taught to read and write and requiring towns with a certain population to build and maintain public schools. The Pilgrim Fathers considered reading to be vital to the success of the colony. It was felt everyone needed to be literate in order to read the Bible, an important force that bound the colony together, and to know and to understand the laws governing the colony (Justice 2013). At first mainly boys attended but eventually all children were schooled.

In 1818 when the Bound Brook Island schoolhouse, overlooking the Herring River, was built there were five schools in Wellfleet. The Island School was an example of a Winter School or Fisherman's School. The economy was based on fishing and most boys over the age of ten went to sea during the summer months. Attendance at the school was in the winter, known as the fisherman's term (Stetson 1963). School was considered very important to these young boys who hoped to become ship captains. Knowledge of arithmetic for business and advanced mathematics for operating navigational instruments such as astrolabes was considered vital.

Nehemiah Somes Hopkins (1860-1953), one of the students at the school, became an eye doctor and later a Methodist medical missionary in Peking (Beijing), China, where he founded a hospital. The plaque on the memorial stone erected on the site of the schoolhouse on Bound Brook Island not far from the Atwood-Higgins property was made in China for the doctor and brought to his former school site (Fig. 8). Probably the most famous student, and Wellfleet's most famous locally born citizen, was Lorenzo Dow Baker. He was born on the island in 1840, a sailor boy at ten, and master of a ship at twenty-one. He was never out of sight of the ocean in his life of nearly seventy years. In 1870 he introduced bananas from Jamaica to the east coast of the United States. In 1878 he founded the Boston Fruit Company, which eventually became the United Fruit Company. Besides his Wellfleet home he maintained a home, hotel, and strong interest in Jamaica. Lorenzo Dow Baker went on to promote tourism in Wellfleet by building the Chequesset Inn on Mercantile Wharf near Mayo Beach in 1885.

With the general collapse in the fishing industry in the 1870s, and decline in population, Wellfleet's schools were being consolidated. In 1880 the Island School was sold. The bell from the school was taken to Jamaica and hung in the Methodist Chapel there.

In 1924 the memorial stone and plaque were erected at the site of the former Bound Brook Island School (Fig. 8). On the plaque was a copy of a poem by former Bound Brook schoolteacher Martha Sparrow Baker:

Tho few remain who once met here
And scattered are afar and near
With love they hold in memory still
The Island Schoolhouse on the hill
And gladly do they mark the spot
That it may never be forgot

Figure 8. Stone erected in 1924 to commemorate the Bound Brook Island School.

Fishermen of the Herring River

When the town of Eastham was founded in 1644, Wellfleet was not yet an independent township. At that time, Wellfleet, referred to as Billingsgate, was considered to be the north precinct of Eastham. Billingsgate extended north as far as Bound Brook, which was called Sapokonish by the Native Americans (Nye 1920). This area was rich in fish and shellfish resources. In 1794 Levi Whitman wrote that "there are few towns so well supplied with fish of all kinds" and that "no part of the world had better oysters" (Whitman 1794, p.119). Billingsgate Harbor became a major fishing area on Cape Cod. The Herring River, as well as several deep-water creeks, emptied into the Harbor, and for many years these locations served as ports which were used by vessels as large as seventy or eighty tons (Whitman 1794). Griffin (earlier called "Griffith's") Island, bordered by the Herring River and Duck Harbor, was particularly well suited for maritime activities; its twelve to fifteen families were sailors and fishermen. The first wharf for Wellfleet's whaling vessels was built sometime before 1720 on the shore of Griffin Island (Deyo 1890; Stetson 1963), probably just seaward of the present Chequessset Neck Road and dike, where the river could be forded at low tide (Stetson 1963). Here sugar, molasses, and "other commodities" (Nye 1920) were off-loaded into ox carts. This "river harbor" served as an active center of mackerel (*Scomber scomrus*) and cod fisheries, with up to 100 vessels ranging from 20 to 50 tons capacity (Pratt 1844).

Settlement reached a peak on Griffin Island, as well as the other islands of the Herring River estuary, in about 1830 (Stetson 1963) and then declined due to rapid shoaling that finally closed Duck Harbor, making it impossible for large vessels to navigate. The last house on Griffin Island burned down in 1890. Settlers left Great Island before 1800 when the last house at Smith's Cove on the Herring River was floated to Dogtown, an area that is now part of South Wellfleet (Nye 1920).

Whaling

Today, the appearance of more than one whale in shallow water is rare, and is categorized as a "mass stranding." In the early days of colonial settlement, stranding of whales on tidal flats was a common phenomenon. Drift whales, i.e., those washed onto the shore, were an economic boon. These beached whales, requiring no fishing effort, were slaughtered and cut up; their oil was extracted from their blubber and heads in "try" works along the shore. The oil, which burned cleanly and without odor, was used before electricity to light lamps. The whale oil was shipped to Boston and earned large profits for the Town and its fishermen. The Town laid claim to any whales floating within a mile of shore (Echeverria 1993). The competition for these whales was so intense that townspeople developed a system that relied on spotters who scanned the Harbor from high vistas on Bound Brook and Great Islands (Braginton-Smith and Oliver 2008) or climbed high poles to look out for whales close to shore. Once the whales, usually pilot whales (*Globicephala melas*) or blackfish as they were called, were spotted, they were surrounded by boats and driven to shore (Fig. 9). The Punonakanits of Billingsgate were hired on the crews of the early whalers and became expert harpooneers. In 1710, the Town voted to appropriate funds so that Joseph Merrick could pay the Native Americans that he hired to assist with his whaling operations on Great Island (Echeverria 1993).

The Billingsgate shore whale fishery, begun as early as the 1640s (Braginton-Smith and Oliver 2008), was centered on Lieutenant Island and Great Island, where housing for fishermen and try works were located. There were also try works on Griffin Island (Nye 1920). The try works relied on wood fires and consumed massive amount of wood. By 1738 the industry had moved to Billingsgate Point (not yet an island), where there was less wood to keep the trying pots boiling, but more whales (Braginton-Smith and Oliver 2008). The stench from the dead whales and trying pots is difficult to imagine. In July 1855, Henry David Thoreau came across a mass stranding of pilot whales that occurred over several days and stretched from Eastham to Truro. He wrote, "About a week afterward, when I came to this shore, it was strewn as far as I could see with a glass, with the carcasses of blackfish stripped of their blubber and their heads cut off…walking on the beach was out of the question on account of the stench" (Thoreau 1865, p. 134).

Pieces of whale bone were found when a tavern site was excavated on Great Island in 1969-1970 by a joint venture of the National Park Service and Plimouth Plantation. In addition to bones, the site revealed parts of harpoons and lances used for butchering whales, and nearby, a piece of scrimshaw, all indicating that the area served as a center for the whaling industry. The tavern was known as Smith's Tavern, after a local whaling family. Most likely, whalemen gathered at the tavern as they waited for whales to appear in the Harbor. There is some speculation that the tavern also served as a brothel. Local legend describes a sign that was placed on the way to the tavern site.

Samuel Smith, he has good flip,
Good toddy if you please,
The way is near and very clear,
T'is just beyond the trees.
(Nye 1920, p.12)

After 1715, shore whaling declined, indicating depleted stocks in Cape Cod Bay that no longer satisfied the demand for whale oil (Braginton-Smith and Oliver 2008); therefore, whaling expanded off shore, with Wellfeet residents traveling great distances for long time periods in pursuit of large whales. By 1771, there were thirty whaling vessels of seventy-five tons in Wellfleet, each employing over a dozen men (Nye 1920). Whale oil became a major commodity of commerce prior to the Revolutionary War (Braginton-Smith and Oliver 2008). It is ironic that whaling was such a lucrative industry in the early history of Wellfleet and the Herring River, but by the 20th century, for health and sanitation reasons, the Town was spending money to bury or otherwise dispose of pilot whales that washed up on shore.

Blackfish Grounding (the beach) - late 1920s

Figure 9. Pilot whale strandings like this were occasions for celebration as residents shared in the profits from the sale of oil rendered from whale carcasses. The photo was apparently taken from Chequesset Neck looking east with the Chequesset Inn on former Mercantile Wharf in the distance.

Inshore fisheries

River herring were seasonally available to the Herring River settlers but many other species of fish abounded near the shore throughout the year. Early on, because the soil could not support farming, Wellfleet citizens turned to local marine resources. Initially, fishing was a sustenance industry with cod, bass, and mackerel constituting the major catch. Cod were caught from boats using hook and line, while seine nets were used to harvest bass, mackerel, and herring (McManamon 1985). When it became possible to preserve fish by salting, and ship it to buyers in Boston and other towns, the inshore fishing industry exploded, transitioning from a sustenance fishery to an export industry. For example, the mackerel fishery started around 1826 and between 1845 and 1864, Wellfleet's mackerel fleet consisted of over 100 vessels (McKenzie, 2011).

Saltworks

Without modern refrigeration fishermen needed a method to preserve freshly caught fish for market. This was done using salt. Thus, the fishing industry created the need for a new industry – the extraction of salt from sea water. The growth of this industry was stimulated by the discovery in the 1770s of the process of solar evaporation and the imposition of a duty on imported salt (Holmes et al. 1995). Salt production became an increasingly lucrative business, and salt works became a common feature on the beaches of Cape Cod. The structures consisted of wooden vats, some as large as 250 by 18 feet, located near the shore and on flat open ground (Fig. 10). Construction required massive amounts of wood. Sea water was pumped using windmills. Often the saltworks were moved and the wood reused (Holmes et al. 1995). Because of their portability it is difficult to know exact locations of saltworks. However, it is recorded that by 1837 there were thirty-nine saltworks operating in Wellfleet producing 10,000

Figure 10. Windmills for pumping sea water into evaporation vats for salt production, mid-1800s. The construction and maintenance of windmills in this harsh coastal environment required great amounts of wood, contributing to local deforestation and eventually forcing the industry to import expensive lumber from off-Cape. The photo shows lumber waste on the ground around these structures.

bushels of salt (Nye 1920). Maps from that period show saltworks on Bound Brook Island (Gillis and Herbster 2013) and specific mention is made of a windmill, most likely used in saltwork operations, south of the Baker-Biddle property on Duck Harbor (Nye 1920; Torp et al. 2013). It is also believed that saltworks existed along the Herring River (Massachusetts Historical Commission 1984). This was a period of great prosperity with fishing at its peak and saltworks up and down the coast. However, by the mid-1800s the salt industry had collapsed. Wood became so scarce that it was brought to Cape Cod from as far away as Maine (Johnson 1997). The salt tariff had been repealed; salt mines had been discovered in nearby states; and costs of production were increasing. No new works were constructed after 1860 and by 1880 the industry in Wellfleet was no longer viable (Nye 1920).

River herring

The Herring River herring run was "established," that is, recognized by an act of the Commonwealth placing responsibility for management with the Town, about 1700 when the Freeman family dug a ditch to Herring Pond (Belding 1920). The river herring belong to two species: alewives (*Alosa pseudoharengus*) and blueback herring (*Alosa aestivalis*). Each spring, these fish migrate up the Herring River to spawning grounds in the freshwater ponds. During their migration, they were herded into a false channel or "pound" (Figs. 3 and 11) a deep horseshoe bend shut off from the main stream at its upper and lower ends by a screen and gate. Fish were caught from the pound in seines and hand nets (Rich 1973) (Figs. 12a and 12b). The gate was closed three days a week by a gate tender, hired by the town, to give the fish free passage to Herring, Higgins, Williams, and Gull Ponds for spawning, and opened for harvest the remaining four days (Belding 1920). Each year, the job of gate tender was offered to the person with the lowest bid for the job.

Around 1800, the right to fish for herring was auctioned off each spring by the selectmen, their authority derived from a vote taken at each Annual Town Meeting. In addition, each Wellfleet citizen was permitted to take 200 fish per year, at a cost of ½ cent each. Local families used the river herring as food and fertilizer. When river herring became important as codfish bait, the town's annual receipt from the auction, ranging between $409 and $667 in the late 1880s and documented in each Annual Report, increased greatly through most of

Figure 11. Fishery station over the Herring River main stem at Bound Brook Island looking up-river, about 1903. Here immigrating herring were diverted into a false channel and netted for food and bait. For bait for the offshore hook-and-line fishery, herring were loaded into barrels, visible on the platform, and transported by railroad; the railroad grade is in the background to the right.

the next decade. At this time proceeds from the sale of the fishing rights were enough to pay the salaries of all elected town officials.

Figure 12a. Dip-netting river herring near Bound Brook Island, 1893. From left to right: Edwin P. Cook, Arthur Cook, Captain Thomas Kennedy.

Figure 12b. Seining river herring in Wellfleet's Herring River, probably early 20th century.

Moving offshore

With increased fishing pressure, the inshore fisheries experienced significant decline. The amount of fish harvested near shore plummeted and thus the cod fishermen of Wellfleet took short trips of a few days to two weeks to Georges Bank. The cod fishing industry shifted from jigging off the boat to long line fishing, and this practice needed bait for hooks. They used "herring" as bait but at the time, herring was a collective term that included mackerel, menhaden (*Brevoortia tyrannus*), alewives and blueback herring. Fishermen relied heavily on river herring (alewives and bluebacks) for fresh bait when the herring were running in the spring, and froze the fish to use as bait during other times of the year (McKenzie 2011). The use of inshore marine resources as bait to support an offshore fishing economy represented a major change in the Wellfleet fishing industry: from a locally consumed harvest to a bait harvest.

Pound net fishery

The fishing industry in Wellfleet experienced a significant change in the mid-1800s when a new style of nearshore fishing was developed. Evolving from the concept of weir fishing, stationary fishing gear was erected along the coast in shallow water. So-called pound nets were composed of a fence-like system of nets that ran from the shore to a pen-like structure set up in deeper waters. The nets were stabilized by vertical poles driven into the sediment (McKenzie 2011). When fishermen closed the pen, the fish were trapped and subsequently loaded onto boats or, if the tide were low enough, put into barrels and carted off by horse and wagon. The pound net fishery proliferated between the mid-1800s and 1930 (Kittredge 1930). In 1889 it must have been quite a sight to see these pounds extended all along the coast of Cape Cod from Sandwich to Truro (True 1887). This system of fishing was very efficient at intercepting large schools of fish. With the increased demand for bait from Banks fishermen, these pounds were very lucrative. After the extension of the railroad to Boston, fish from the pounds were packed in ice, shipped by train, and sold to fresh fish markets. Whether used for bait, fertilizer, or food, great quantities of river herring, menhaden, mackerel, bluefish (*Pomatomus saltatrix*), striped bass, and other species of fish were harvested from the sea.

The pounds caused the demise of the inshore fishery that took fish, one by one, on bait and hook. The large volumes of fish taken by the pounds glutted the market and caused the price of fish to drop. Furthermore, the traditional hook and line fishermen were harvesting fewer and fewer fish. By the late 1800s, even the pound fishermen were experiencing major declines in catch, and eventually, the inshore fishery collapsed altogether.

Methodism and the Herring River valley

In the early 1800s when the Herring River valley was thriving with a maritime economy, it also experienced the Second Great Awakening, an evangelical movement that was sweeping America. Methodism was the religion that took hold in Wellfleet. Religious zeal was particularly strong on Bound Brook Island. In an account of Methodism in Wellfleet written in 1877 mention is made of 'Brown Brook', which is most certainly Bound Brook. On that island "scarcely an adult is left unconverted and not a single family but some of which have found a pardoning God" (Palmer 1877, p. 30). Among the many Wellfleet citizens who in 1816 were 'born of the Spirit' was Joel Atwood (Palmer 1877). Atwood lived in a home he built on Bound Brook Island (Historic American Buildings n.d.) and seems to have been especially affected by his Methodist faith. "Brother Joel Atwood was so full of love, that one day, some little time after his conversion, while at work on the roof of his house, he was heard by his neighbors repeatedly to shout, 'Glory to God!' 'Glory to God!' and then he would sing one of the good old hymns" (Palmer 1877 p. 23).

During the Second Great Awakening, Methodists relied on traveling ministers to bring the faith to people in remote areas. Most of these ministers were powerful orators who could move crowds to a frenzy. One of these was Lorenzo Dow who traveled extensively throughout the United States. He was so well known at the time and so inspired his audiences that hundreds of boys were named for him (Price 1912). It is almost certain that the Baker family of Bound Brook Island, who were fervent Methodists, named Lorenzo Dow Baker, Wellfleet's most famous native son, after him.

Bound Brook Island was also the site of Methodist camp meetings from 1823 to 1825. Here preachers, one after the other, emotionally moved crowds to cry out, sing, and call each other to repent. Sermons 'awakening' people to God were the main purpose of these camp meetings, but they also were a diversion from daily hard work – fishing, farming, husbandry, housekeeping- and provided a forum for community gatherings (Hurd 1884).

Chapter 4. Resource depletion, 1600-1800

By the mid-1800s the residents of outer Cape Cod were faced with the bleak consequences of the depletion of once-abundant native natural resources. Nearshore fisheries, ranging from large whales to small forage fish such as mackerel, had been fished out, massive oyster reefs had been harvested and mined for lime production, and most of the forest had been cut to build houses, ships, and saltworks, and to fuel homes and try works where whale blubber was rendered into oil.

Marine resources

It was perhaps a sign that the number of whales found on the shore was declining when, as early as 1707, an attempt was made to regulate the nearshore whaling industry by limiting the enterprise to locals, i.e., residents from Great Island (and also Lieutenant Island). The Town demanded two shillings from the harpooneer or steersman of every boat that had an "outsider" on its crew (McManamon 1985; Braginton-Smith and Oliver 2008). The enforcement of this regulation was challenging, so it was abandoned in 1712 at Town Meeting (Braginton-Smith and Oliver 2008). Nearshore fisheries for mackerel, cod, and other fish originally supplied food and bait for the townspeople, but when it became possible to salt and preserve fish, or to ice them and send them to Boston by railroad, more and more fish were harvested and ended up in urban markets. As the commercial exploitation of fish increased, fish stocks decreased and the local fishing economy involved trips farther and farther from shore.

Oysters, including those from the Herring River, were a valued commodity. But, by around 1770, there were no oysters left in Wellfleet. Over-harvest is one of the reasons for the decline of the oyster industry, but there were also other factors at play. As oysters were harvested, their shells were removed from the ecosystem. The shells of old oysters are necessary substrates for the settlement of "spat" or oyster larvae. Without suitable substrate, the oyster larvae are buried by sediment and fail to develop. Already compromised, the remaining stocks in the Harbor were most likely wiped out by an oyster disease. In response to this latter oyster crisis, oysters were imported from other parts of Massachusetts such as Taunton and Wareham and from Chesapeake Bay. These efforts allowed the commercial oyster industry to regenerate (McManamon 1985).

Deforestation

The effects of deforestation on upland soils first became apparent in 1740 (Echeverria 1993). Trees were cut for building and heating homes, fueling try pots and saltworks, and shipbuilding. Grazing by domestic animals also took a toll on the land. Pratt, in 1844, reported that "no wood is left in the township...except a tract of oaks and pines, adjoining the south line of Wellfleet" (Pratt 1844, p. 4). As early as 1660, there were 7000 to 10,000 sheep on the Cape, mostly in Eastham, Billingsgate (the Eastham precinct and future Wellfleet), and Truro (Echeverria 1993). Unsustainable deforestation and intensive sheep grazing left sandy upland soils exposed and "broken" (disturbed and unstable) (Fig. 13). Livestock were often confined to coastal islands, to protect them from predators (McManamon, 1985). As early as the mid-1700s, acts had been passed by the General Court to "Prevent Damage Being Done unto Billingsgate Bay (Wellfleet Harbor) in the Town of Eastham by Cattle, Horse kind, and Sheep Feeding on the Beach and Islands Adjoining Thereto..." in 1742 and to "Prevent damage being done on Bound Brook Island, and Griffith's (now Griffin) Island, within the District of Wellfleet..." in 1768 (Echeverria 1993, p. 109). In 1801 Town Meeting approved a petition "... to prevent damage being done to Wellfleet Harbour (sic) by excessive numbers of cattle, horses and sheep going and feeding on the island beaches southward of Griffin's Island on the west side of said

Figure 13. Bound Brook Island largely devoid of trees in 1900. The cutting of trees for fuel, as well as for house, ship, and saltworks construction, eliminated most of the outer Cape forest by 1850. The forest has since recovered over most of the landscape in this photo.

Harbour." Despite similar restrictions voted in 1810, 1816, 1822, 1833, and 1873 (Town Meeting records), the thin native topsoil eroded into already shoaling creeks and bays, and shifting sands filled formerly important harbors at Duck Harbor and the mouth of the Herring River, rendering them unnavigable (Deyo 1890).

Chapter 5. Mosquitoes and the diking of the River, late 1800s -1909

Mosquitoes have long been a nuisance on Cape Cod; Henry David Thoreau (1865) complained about them on his 1850s hikes across the Cape. Despite dire accounts of the abundance and voraciousness of these pests near Herring River, no historic records of the specific mosquito species involved have been found; however, modern surveillance indicates that the most numerous nuisance mosquito species currently emerge, and probably emerged, from coastal marshes. These are floodwater-breeding species, which deposit their eggs in depressions on wetland surfaces that subsequently flood with rainwater or tidal waters; the eggs hatch and mature to flying adults, provided the larval mosquitoes survive normally intense predation, particularly by estuarine fish. These floodwater mosquitoes include species whose larvae can survive in fresh, brackish, or salty water.

Late 19th century attempts at mosquito control employed small drainage projects to dewater breeding sites, and the application of kerosene to the water surface to suffocate larvae; however, during rainy years the nuisance persisted. Proposals for mosquito suppression appeared first in the 1904 Wellfleet Town Annual Report, with a request for $1000 "to drain and dyke (sic) meadows and use oil where needed to stop the mosquito pest." Some salt marshes were already tide-restricted by dikes by 1900, probably to increase pasturage and tillable land and ease seasonal salt hay harvest, e.g., Pole Dike on former Doane's Run, now Pole Dike Creek, draining about 200 acres of original tidal marsh upstream of Pole Dike Road. Additionally in 1905 a Town Meeting article was passed requiring the owners of "salt and fresh meadows to cut ditches" and connect them to main creeks; landowners who failed to comply would be charged for ditching undertaken or contracted by the Town. The first mention in Town Meeting records of diking Herring River occurred in 1905, with a vote to "appoint a committee and petition the Legislature to build a dyke (sic) across Herring River."

Until that time most of the tidal flow in and out of Wellfleet's bayside salt marshes remained unrestricted. Even the railway, constructed about 1870, employed bridges rather than solid-fill dikes to pass over Pole Dike Creek and the main stem of Herring River.

Status of the herring run

By the late 1800s, the harvest of some natural resources from estuaries like the Herring River was still an economic and social focus of the town. An important example is the consistent attention given to the management of the herring run in Annual Town Reports. The Treasurer's Report of 1883 included a "Brook Account" listing all wages associated with operation of the run. A "Fishery Station," where herring were trapped and netted, was staffed daily throughout the spring run. An Annual Town Meeting vote in 1896 was required to move the "Fishery Station from its present location (perhaps the vicinity of High Toss Road) to the Bridge at the southeast end of Bound Brook Island" (most probably where modern Bound Brook Island Road crosses the river). These river herring migrated the full length of Herring River each spring to spawn in all four headwater ponds, Gull, Herring, Higgins, and Williams, and likely also traversed Pole Dike Creek to spawn in Perch Pond (Rich ca. 1971).

Despite the pressures of the coastal weir fishery (True 1887, McKenzie 2011) and annual netting of spawners in the river itself, over 200,000 fish were harvested annually between 1888 and 1890, the only years for which total catch was reported. Dr. David Belding, in his 1920 *Report on the Alewife Fisheries of Massachusetts* said that Wellfleet's Herring River run was the second most productive in the state between 1890 and 1899. In 1902, the Gloucester Mackerel

Company won the Wellfleet river herring auction with a bid of $1310.51 (Annual Town Report).

Motivated apparently by the success of the run and auction, the Town voted to extend the herring spawning area in the headwater kettle ponds by constructing a "Gull Pond Canal," known as the "Sluiceway," connecting Gull and Higgins Ponds in 1903 at a cost of $569. [Note, however, that Belding (1920) reported that the connection between these two ponds was dug in 1893; moreover, recent paleo-limnological work by Winkler and Sanford (1994), who analyzed zooplankton remains in Gull Pond sediments, indicated that these two ponds had been connected, likely by Native Americans, hundreds of years earlier.] The Sluiceway continued to fill in periodically with sediment, however. In 1946, another town vote appropriated $500 for a permanent Sluiceway, which wasn't very permanent. In 1956, the Wellfleet Highway Surveyor reported that the Sluiceway was cleaned out four times, while in 1958, the town voted to use $500 for the rebuilding of the Sluiceway by Wellfleet Boy Scouts of America, Troop 81, "under competent supervision" (Town of Wellfleet, Annual Report).

Interestingly, the Annual Town Meeting of February 1906 broke with tradition and voted to leave the "gate of Herring Brook open and not sell the fishing of the Brook." This may have been in anticipation of diking the river, first proposed in 1905, and concern for its effects on herring migration. However, the following month a Special Town Meeting reconsidered and reversed this vote. There is no explanation of either action in the Annual Report, but one might suspect public ambivalence about disturbing a traditional and still productive fishery.

Planning for a dike across the Herring River

By 1906 Lorenzo Dow Baker (Fig. 14), the local sea captain, entrepreneur, and town philanthropist, who had become wealthy importing bananas, began campaigning for the diking of the Herring River and other Wellfleet Harbor estuaries to drain and eliminate mosquito breeding sites. Baker was already involved in the budding tourist industry, having built the large and opulent Chequesset Inn near the river mouth in 1885. He also owned land within the Herring River flood plain and promoted the diking for future cranberry farming, in addition to mosquito control. Importantly, up until this time only upper Pole Dike Creek marshlands, between Pole Dike Road and the current Route 6, had been tide-restricted; the railway built across the flood plain about 1870 employed wide trestles to cross the river, with little impedance of tidal flow.

A town committee headed by Baker solicited a proposal from Whitman and Howard Engineers for an assessment of the potential of diking the river and for engineering design (Whitman and Howard 1906). The optimistic objectives are excerpted below (Fig. 15).

Also in 1906 Baker's committee petitioned the Commonwealth to enact legislation and to partially fund the dike. Accordingly, Chapter 400, Section 1 of the Acts and Resolves of the Massachusetts Legislature of 1906 authorized the diking of the river "provided that the dike shall contain a proper fishway which shall be approved in writing by the commissioners of fish and game." Section 2 of the act stated that it would take effect "upon its acceptance by two-thirds of the voters of the town...at a special meeting called for the purpose." In 1907 the town voted $10,000 to dike the Herring River, provided that the state match that amount of funding.

Figure 14. Captain Lorenzo Dow Baker, Wellfleet fisherman, importer of bananas, coastal developer and philanthropist. Baker argued that diking Herring River and other Wellfleet tidal marshes would reduce mosquitoes and yield more land for seaside development, stimulating the new tourist economy. The town decided to build the dike in part to honor locally popular L.D. Baker shortly after his death.

In April 1908 the Board of Harbor and Land Commissioners, under authority received from the Massachusetts General Court (Legislature), wrote to the US Army Corp of Engineers (ACOE) for permission to "construct a dam, with sluiceways therein, at the mouth of Herring River in the town of Wellfleet, for the purpose of excluding the tide-water from said river and draining the marshes" (Army Corp of Engineers 1908). On 10 June 1908 the ACOE held a hearing in Wellfleet to solicit comments from the public. The one stated ACOE purpose of the hearing was to determine if the river was still navigable before permitting the dike. Many local boat captains testified that the river was no longer navigable. However, Levi Higgins, a retired fisherman and current salt hay farmer, argued that the river was still navigable by skiff and opposed the dike; he later threatened a suit against the town for the dike's destruction of his salt hay crop. At that same meeting, the chairman of the Massachusetts Fish and Game Commission testified that the dike would benefit shellfish by increasing "nitrification" and the food supply for shellfish from the diked and freshened, versus existing brackish, marshes (Army Corp of Engineers 1908).

With workers and material already gathering at the future dike site, Town Meeting convened in the fall of 1908 to consider blocking the tidewater from Herring River. Proponents listed three benefits: 1) improvement in fisheries, 2) development of cranberry bogs, and 3) mosquito control. Arguments for and against the project were roughly equal, until it was suggested that the project would honor recently deceased Lorenzo Dow Baker who had given so much to the town during his lifetime. This sentiment apparently carried the day, as the vote was 188 to 40 to build the dike.

The dike between Chequesset Neck and Griffin Island was constructed during the winter of 1908-1909 (Figs. 16 and 17). About the same time, a smaller earthen dike was built across the remaining, albeit shoaling, inlet connecting Duck Harbor to Cape Cod Bay. After the dikes were constructed, their ownership and maintenance became the responsibility of the Town.

> BOSTON, February 5, 1906.
> MESSRS. N. H. PAYNE, FREDERIC W. SNOW, M. D. HOLBROOK, THOMAS A. NEWCOMB, L. D. BAKER, JR.,
> *Committee*, Wellfleet, Mass.
>
> *Gentlemen,*— We have been asked by you to make surveys, plans, and estimates, and report on the matter of diking out and draining the large area of marshes tributary to Herring River in the Town of Wellfleet. We have caused such surveys and examinations of the locality to be made as the short time available will permit, and we think we understand the situation. The problem is one of considerable magnitude; but, after having given it due study, we believe it can be solved without undue expense, and all the underlying principles of its accomplishment are, after all, simple as to fundamentals.
>
> We understand the first and main object sought is to exterminate the mosquito pest; the second, the draining of the marshes so they may be brought into valuable lands; the third, to transform the unsightly swamps (which must, as they grow worse and more neglected, be a direct menace to health) into clean and healthy areas which will add to, instead of detract from, the beautiful landscape with which nature has richly endowed this locality. We do not consider ourselves authority on the very aggravating mosquito problem; but, from a smattering of information,— which is all we can claim as to this particular phase of the matter,— we have no doubt that you are proceeding in the right direction, and accomplished in this way the work is done for all time.

Figure 15. Excerpt from opening page of the 1906 Whitman & Howard Report to the Town of Wellfleet on the Proposed Dike at Herring River, outlining the objectives of the project.

Figure 16. Ceremonial start of construction of the Chequesset Neck dike, August 1908. Curiously, the project was not passed by Town Meeting until the fall of that year, according to Annual Reports.

Figure 17. Construction of the Chequesset Neck dike across the mouth of Herring River, 1908-1909. As seen in the photo above and the photo at the top of the following page, the dike was built without using coffer dams to block river flow, meaning that workers had to fight the flooding and ebbing tides. The causeway was constructed with soil dug from borrow pits at either end of the structure and carried on carts across the dike on a rail system. The embankment was stabilized with shipped-in stone and locally excavated marsh peat. The third photo (bottom of next page) shows the newly completed dike.

Figure 17 continued…

Chapter 6. The diked Herring River, 1909-1960s

The dike's effects on financial and natural resources, and mosquitoes

Although significant amounts of money were spent by Wellfleet on ditch drainage for mosquito suppression shortly before the diking, these expenditures increased greatly after the intended mosquito-control dikes were already in place (Fig. 18). Suppression included ditching and stream "cleaning" (channelization), as well as the application of oil (kerosene) to smother and poison aquatic immature mosquitoes. Through the 1910s and 1920s annual expenditures for mosquito control regularly exceeded $1000 (over $24,000 in modern currency). Between the diking of Herring River in 1909 and 1935, when mosquito control began to shift from the town to the state, albeit at continuing town expense, Wellfleet spent over $31,000 (over $500,000 in 2009 dollars) on ditching and oiling diked wetlands. This expense was in addition to the cost of constructing and maintaining the many dikes built in town since 1910.

The wording of many subsequent Annual and Special Town Meeting appropriations for the drainage of diked marshes suggests that the predicted outcome of blocking the tides, i.e., fewer mosquitoes, was not being realized. A typical example is Article 17 of the Annual Town Meeting of 1910, the year after the River was diked:

*To see what action the Town will take towards raising and appropriating a sum of money for the purpose of establishing proper outlets for the water in the different sections of the Meadow lands, which **are affected by the Herring River Dike**... (*emphasis ours).

Despite the effect of the dikes in impounding water, reducing tidal flushing and thereby favoring floodwater mosquito breeding, local officials and Town Meeting chose to continue with tidal restrictions and ditch drainage, rather than to restore natural tidal flushing.

Figure 18 shows the collapse in the herring fishery after the diking of Herring River (note that Belding (1920) attributed this collapse to the one-year lease system), but other natural resources of economic value were lost. In 1910, the herring auction only brought in $70 to the Town's coffers, and after that, there was no auction because there was "nothing left to sell" (Sterling 1976, p. 12). In 1913, Town Meeting appointed a committee to arbitrate the case of damage done to Levi Higgins's meadow, caused by the Herring River Dike; this man was apparently harvesting salt hay that, with the blockage of sea water, would have succumbed to the invasion of freshwater wetland plants. On another occasion, and as late as 1927, Annual Town Meeting voted that "land damages on Herring River Meadows be left in the hand of the Selectmen for adjustment," presumably financial compensation for continuing damage to salt hay crops. As mentioned, the expansive salt marshes and abundant salt hay for animal fodder was the original attraction for the Pilgrims of Plymouth to relocate to the outer Cape (Cumbler 2014); Rich (1973) reported that the Herring River marshes produced winter fodder for 200 cattle around 1893, shortly before the system was diked.

The blockage of tides in 1908-9 opened the way for further alterations of the flood plain. During the 1920s, a plan to further drain the wetlands upstream of the Chequesset Neck dike comprised deepening, straightening, and channelizing the main stem from High Toss Road to the present Route 6 (Fig. 19), and replacing the bridge at High Toss Road with a narrow (five-foot-diameter) culvert (Massachusetts Department of Public Works 1923). (Note that the culvert was not installed apparently until 1958; see below.) Natural meanders were

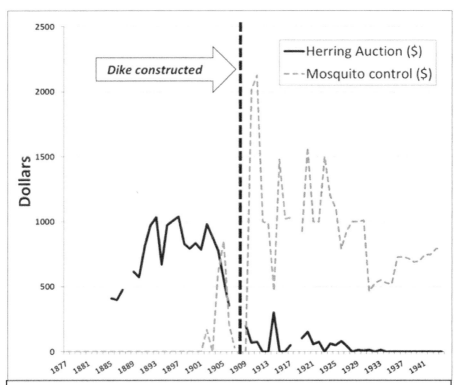

Figure 18. Proceeds from the annual herring run auction versus expenditures for mosquito control, 1877-1940s. The Herring River run, already in decline due to overfishing (Belding 1920), collapsed after the river was diked. Mosquito control expenses, including ditch digging, river channelization, and the application of pesticides, increased greatly. Source: Wellfleet Annual Reports.

isolated from flow as the main stream was straightened; excavated spoil was dumped along the banks in large piles forming elevated levees, which probably impeded water flow onto and off of the interior marsh during periods of high water. With slower water flow, parts of the Herring River began to fill in with sediment, requiring frequent maintenance. Many Town Meeting reports beginning in 1917 tell a story of votes taken to appropriate funds to "clean" the Herring River, Pole Dike Creek, and other branches of the River. The task of the

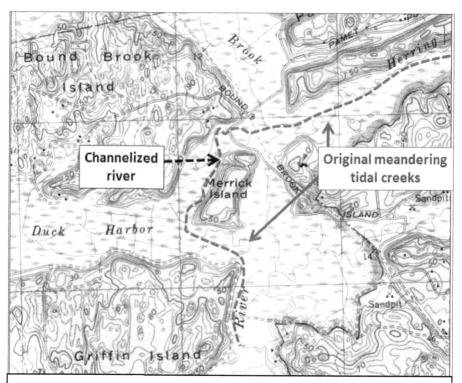

Figure 19. To further a program of wetland drainage, the Herring River main stem was channelized and straightened from High Toss Road to Bound Brook Island in the 1920s (Massachusetts Department of Public Works 1923). This project was apparently continued in the 1930s, cutting off meanders all the way to the present Route 6 and leaving high piles of dredged material, now covered by upland plants and large trees, on the river bank.

annual channelization and raking of aquatic vegetation from the river from High Toss Road to Route 6 was taken over by the Cape Cod Mosquito Control Project (CCMCP), probably in the late 1930s; this work was voluntarily terminated by the project in 1984 amid concerns for sediment suspension, dissolved oxygen depletions, and juvenile river herring die-offs.

Early issues with the dike

Town meeting reports record the many votes taken to make repairs to the dike, for example, in 1915, 1919, 1921, and 1922. In 1930, The Wellfleet Board of Selectmen wrote in its annual report, "The condition of the Herring River Dyke (sic) has been the cause of considerable anxiety and concern on the part of your Board during the past year, and considerable work has been planned to be done on the west side of the structure, which we sincerely hope will improve and continue to preserve its current condition of safety and permanence." In 1933, Fish Constable Henry DeLory's Annual Report states, "Early last spring, a bill was introduced into the legislature for the repair of the Herring River Dyke (sic) and the sum of $10,000 was appropriated for this purpose." Wellfleet's share was $108.75. He further writes that the repairs that were done were "highly unsatisfactory."

The roadway over the dike was widened in the mid-1930s and guard rails were installed. In addition, a three-acre parking area was constructed on Griffin Island " extending from the shores of Cape Cod Bay to the Shores of Wellfleet Harbor...thereby providing for a scenic highway for the Town of Wellfleet unsurpassed by any on Cape Cod" (Annual Report 1934). By 1934 High Toss Bridge was falling apart and the Town rebuilt it. The Lorenzo Dow Baker estate donated fill, which allowed the bridge to be shortened from 106 feet to 10 feet, and thus require less maintenance. High Toss Bridge Road was scraped and filled again in 1935. In the mid-1940s, the dike was damaged by a severe storm and temporary repairs were made. By 1957, High Toss Bridge was seriously deteriorating and in 1958 was replaced by a five-foot-diameter culvert.

About 1929 a golf course was built by the Chequesset Yacht and Country Club in the Mill Creek sub-basin of Herring River, just above Chequesset Neck Road; five of its nine fairways were established directly on the original tidal marsh surface, apparently without fill to raise ground-surface elevations. In 1951 the new Route 6 was under construction through north Wellfleet and needed topsoil for side dressing; much of this topsoil was obtained by scraping tens of acres of Herring River marshes between High Toss Road and lower Pole Dike Creek (Oscar Doane, Cape Cod Mosquito Control Project, personal communication 1982, Fig. 20), showing the low value placed on wetlands even at that late date. In the late 1950s, two homes were built at very low elevations on the river bank between the dike and High Toss Road.

Figure 20. Aerial view of lower Pole Dike Creek marshes after being scraped for topsoil to side-dress the new Route 6, under construction in 1951. Densely spaced tracks of the heavy equipment are evident in this winter scene looking south. High Toss Road skirted the left (east) side of the flood plain, as it does today. Note that the trees and shrubs that now cover most of the flood plain had not yet established by the 1950s. Photo courtesy of the Cape Cod Mosquito Control Project.

The Cape Cod National Seashore, a unit of the National Park Service, was established in 1961, incorporating about 80% of the Herring River flood plain within its boundaries; however, the National Park Service did not receive ownership and/or control of the dike that exerted so much influence on the hundreds of acres of diked wetlands upstream. Although most wetland encroachment by development within the Seashore ended, it continued around the Herring River sub-basins outside the Seashore, near Mill Creek and Upper Pole Dike Creek (Fig. 21). Passage of the Massachusetts Wetlands Protection Act in 1972, and its enforcement by the local Conservation Commission, further limited wetland encroachment both within and outside Seashore boundaries.

Meanwhile, effective blockage of tides caused wetland water levels, and surface-water salinity, to decrease. The freshening and drainage enabled freshwater wetland grasses, forbs, and shrubs to invade and out-compete native saltmarsh grasses. Much of the original salt marsh in the lower river between the dike and High Toss Road was replaced by cattail, which persisted into the early 1970s (Snow 1974). Over higher portions of the flood plain, e.g., between High Toss and Old County Roads, the peat was perennially drained, allowing the growth of upland trees, vines, and shrubs including many invasive and non-native species (Portnoy & Soukup 1982).

In addition to affecting the herring run, the dike also had a major effect on other fish, birds, and turtles. Above the dike, shellfish disappeared. With the loss of the salt marsh, important foraging and nursery habitat for finfish was eliminated. The feeding ground for American black ducks (*Anas rubripes*) and brants (*Branta bernicla*) diminished, while nursery, foraging, and nesting habitat for diamondback terrapins (*Malaclemys terrapin*) was eradicated.

Chapter 7. Shift to tourism and summer residents, late 1800s-1950s

While the Town debated the issue of diking the Herring River, a new economy was taking hold: it was based on tourism and summer residents. Prior to 1840 the River Wharf Company maintained its wharf, packing house, and "fitting-out store" (ship's chandler) in the "apparent security of the harbor at the mouth of Herring River" (Deyo 1890); this broad reach of the river had earlier been called "the great Basse (sic) Pound" (Echeverria 1993). In short, the shallow bays and tidal creeks had been for 250 years not only harborage, but also the principal routes of transportation, commerce and communication among the town's main settlements at Bound Brook Island, Griffin Island, Coles Neck, and Duck Creek (Fig. 21). Although there was enough water at high tide in 1800 to float Reuben Rich's recently constructed schooner *Freemason* (of "100 tons burden," probably about 100 feet long) from the boatyard at the southeast corner of Bound Brook Island to the Bay (Nye 1920), by late in the century the creeks of the Herring River system were barely navigable by skiff (Army Corp of Engineers 1908).

This loss of navigability, along with diminished fish and shellfish, further reduced the estuary's value in the eyes of the local community. Increasingly, roads and bridges were constructed for overland travel among the settlements around the Herring River (Fig. 22). The railroad was built across the flood plain in the late 1860s. There was little left to sustain the fishing and farming community on the north Wellfleet islands. After 1830 most families relocated to the shores of Duck Creek, bringing along their disarticulated ("flaked") houses on barges from Griffin and Bound Brook Islands and reassembling them in the new town center. Many other people left the Cape entirely. By 1850 Duck Harbor was no longer navigable, and only five houses were left on Bound Brook Island (Rich 1978).

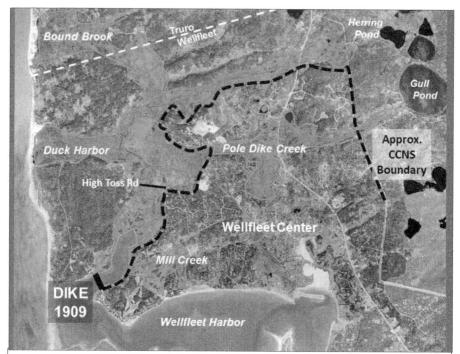

Figure 21. The present Herring River flood plain, its sub-basins and the Cape Cod National Seashore boundary. As indicated, Mill Creek and upper Pole Dike Creek sub-basins are outside the Seashore.

The few families remaining in Wellfleet in the late 1800s increasingly shifted their economic focus from traditional resource extraction to development for tourism. The arrival of the railroad in 1869 improved overland access for the vacationers. People could now use the train, and later improved roads, to escape the city for a day or two or for the entire summer season. The attraction of the Town's beaches, ponds and rural atmosphere to vacationers and summer residents was apparent: by 1920 the Wellfleet Board of Trade was advertising the Town as a tourist destination.

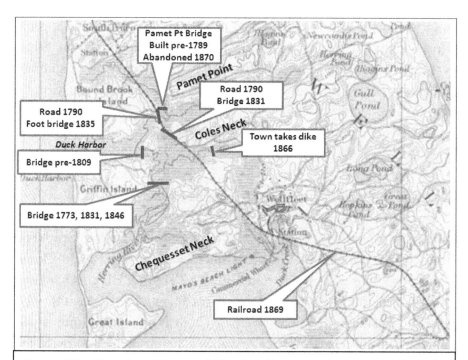

Figure 22. Historic crossings of Herring River marshes and tidal creeks interpreted from Wellfleet Town Meeting actions reported in Annual Reports. Bridges were constructed to allow overland passage among settlements on Griffin Island, Bound Brook Island, Coles Neck, Pamet Point, and around Duck Creek to the east. The latter would become Wellfleet Center as the outlying islands and necks were abandoned following the local depletion of natural resources including the shoaling of creeks and harbors through the mid-1800s. The base map is the 1887 US Geological Survey Wellfleet Sheet.

Wellfleet is unsurpassed as a vacationland. To those who enjoy the beach with its pure clean sand and warm bathing in clear water, or a cold plunge in the rugged surf...You will find our hotels and inns offer you quiet and restful comfort under perfect conditions... (Donaldson et al. 2010, p. 31).

Thus began a new chapter and a new economy in the history of the Herring River valley; one centered on Bound Brook Island. In the 20th century, Bound Brook Island had a number of unique and often colorful individuals who were drawn to its relative isolation. Some were full-time residents and others introduced the concept of summer residents. They included diplomats, artists, writers, and the world's greatest mid-century architects. Here they created a community of intellectuals. As one of them expressed it, he had found nirvana (Kino 2014).

George Higgins and the Atwood-Higgins House

In 1919 George Higgins, a businessman from Brookline Massachusetts, inherited his great uncle's home on Bound Brook Island. The house was built around 1730 by Thomas Higgins who was a fourth generation descendant of an early settler in Plymouth who moved to Eastham. Thomas originally built it as a half Cape; it was enlarged to a full Cape by later Atwoods. The last person to make the Atwood-Higgins House his permanent home was Thomas Atwood, Jr. who died in 1873. During the nearly fifty years that he lived there, he saw Bound Brook Island at its economic peak but also saw its desertion by residents due to the loss of the fishing industry. Atwood Jr.'s children, including Captain Edward B. Atwood, inherited the property but seldom visited it. As a result the house and property were mostly uninhabited for forty-six years. In 1919 George Higgins, nephew of Captain Atwood, purchased part interest in the property with the stipulation that he maintain it in his uncle's absence. In 1920 the Captain died and George Higgins inherited the property. He took great interest in and personal responsibility for renovating the property. He spent nearly forty years dealing with poor road access, vandalism, storms, and constant maintenance (Burke 2009). Using local labor he also added a number of other structures, which fit his romantic idea of how a colonial-era village should look. He personally designed and oversaw the construction of his village and the landscaping of the property. One of the buildings he added was a country store which served as his office and as a place to entertain guests (Fig. 23). As part of his fondness for the past he modeled this building on one he visited as a child with his grandfather in Vermont.

George Higgins's love for his summer home and Bound Brook Island reflected the change in the attitude toward the remote outer Cape. With the railroad, automobiles, and better roads a leisure industry developed as a major economic activity. While George Higgins valued his privacy and took steps to keep his property relatively inaccessible, he did welcome visitors. In 1951, at least 135 people came to his country store as members of tours led by the Wellfleet Historical Society and Museum (Donaldson et al. 2010).

In 1961 the Atwood-Higgins House, outbuildings, and 50 acres were donated by George Higgins to the newly created Cape Cod National Seashore (Seashore). According to his wishes the property was named after both families - Atwood and Higgins - because of the 156-year ownership of the property by the Atwoods.

Francis and Katherine Biddle and the Baker-Biddle property

The history of the Baker-Biddle property in many ways parallels that of the Atwood-Higgins House located about a mile away. These properties were inhabited for several generations by fishermen, and afterwards abandoned by their owners in the mid-1870s due to the economic decline in the fishing industry. They were then resurrected by summer residents who came to enjoy the peace and relative seclusion of the area in the early 20th century. The Baker-Biddle property is named after two families who lived there. The first is the Baker family. David Baker, Jr. was the builder of the original 1790s house, and it was generations later the birthplace in 1840 of Lorenzo Dow Baker.

Figure 23. George Higgins's New England country store and post office. George Higgins built a number of outbuildings to create his romantic version of a colonial era village, including the country store and post office. In the 1950s it was a favorite destination for tours offered by the Wellfleet Historical Society and Museum. Photo courtesy of National Park Service.

Katy Dos Passos, the wife of the author John Dos Passos bought the 180-acre property from the Bakers in the 1930s for $3500. She owned it for a very brief time, but the Dos Passos name associated with Bound Brook Island heralded a new era when a number of artists and intellectuals would live in the area. The house and surrounding acreage was sold in 1937 at the same price to John Hall, a self-taught modernist architect and a key figure in the modern history of Bound Brook Island.

Like George Higgins, his neighbor on Bound Brook Island, Jack Hall began creating a small estate. From 1937 until 1949 he modernized the main house and made additions to it. He rebuilt the old barn from recycled materials using wood that came ashore from nearby Cape Cod Bay. He collected this with an old Ford station wagon and later a horse (Hall 2003). Jack Hall used the barn to house small farm animals – his idea of a country farm. The oldest building on the property dates from around 1690 and was moved from either Great Island or Provincetown where it had originally served as a try work. It was nicknamed the Delight Cottage after a sign affixed to its exterior. The sign was taken from a brothel in Provincetown. The last building that Jack Hall installed on the property was an old barn he bought from a neighbor and moved to his property. This later served as a cottage. This reconstruction work helped crystallized Jack Hall's idea of architecture. He used materials, many recycled, which showed a sensitive appreciation for the history and appearance of the landscape. This philosophy would later be apparent in his design of Modernist houses.

The Biddles are the second family whose name is now associated with the property. In 1949 Jack Hall sold the Baker estate and 10 acres of land to Francis and Katherine Biddle who had summered there the previous year. This couple attracted intellectuals to their summer retreat on Bound Brook Island. Francis was introduced to the outer Cape by his artist brother, George Biddle, who had bought a home in Truro in the early 1940s. He and his wife, Belgian sculptress Helene Sardeau, were among the distinguished artists who were drawn to the outer Cape. The Biddles are a prominent Pennsylvania family. Francis Biddle (1886-1968), the new owner of the property, was a great supporter of the New Deal and a defender of labor. He served as

Franklin Delano Roosevelt's Attorney General during World War II and after that war became the chief American judge at the Nuremburg trials. He wrote a number of books including the 1942 biography of Supreme Court Justice Oliver Wendell Holmes, *Mr. Justice Holmes*.

His wife Katherine Garrison Chapin Biddle (1890-1977) was equally accomplished. She was a poet, critic, and civil rights activist. Her poem the *Lament for the Stolen* which was a response to the Lindberg baby kidnapping was set to music by Eugene Ormandy. Her poem *And They Lynched Him from a Tree* was set to music by William Grant Still (1875-1978), the "Dean of Afro-American Composers" and performed in 1940 by the New York Philharmonic. Archibald MacLeish, friend of the Biddles, appointed her as one of the original Fellows in American Letters of the Library of Congress, and she served as a judge of poetry for a number of awards including the prestigious Yale University Bollingen Prize in Poetry. She was a poet's poet and spent several weeks each year on the lecture circuit promoting poetry (Rigolot 2011).

The Biddles entertained extensively. In her diary Katherine specifically mentions Conrad Aiken, Arthur Schlesinger Jr., Archibald MacLeish, the Nobel Prize poet Saint John Perse, and Edmund Wilson, another Wellfleet resident (Rigolot 2011). It was a literary group and they discussed poetry, often having readings. Francis Biddle remarked on their social life in Wellfleet.

There are I suppose as many cocktail parties on the Cape as in Washington.....At our end of the Cape one rubs elbows with musicians and painters and writers. Xavier Gonzales, Gardner Jencks, Edward Dickinson, Ed O'Conner, Edmund Wilson, Waldo Frank, and Arthur Schlesinger, Jr.....and a long haired pretty beatnik now and then, with overalls and dirty feet, whose mission is to remind us older birds not to be so Goddamn fussy (Biddle 1962, p. 487).

Francis died of a heart attack in Wellfleet in 1968 and Katherine died in 1977. In June 2011 the Trust for Public Land bought the property from the Biddle family for the National Park Service at a cost of over two million dollars. There are various agreements between the National Park Service and owners of land within the Cape Cod National Seashore as to sale of land and land usage. This piece, which consists of 10 acres, could have been sold and a number of large vacation homes built. When the idea of the Cape Cod National Seashore was emerging Francis at first was in opposition to it. However, he later embraced the idea writing:

The law creating the park on the lower Cape was adopted a year ago, and our house is included. Perhaps the dunes with their delicate lines and tracery can be preserved from the careless tramp of the vacation crowd. Perhaps we can teach our children reverence toward the world of nature, yielding to her natural right. I do not wish to see the lower Cape turned into a suburb of Boston. The long problem of the future is the relation of swarming mankind to this little planet that they have overrun and despoiled (Biddle 1962, p. 487).

Jack Hall and the Modernist Architects

John "Jack" Hughes Hall graduated from Princeton University in 1935 with a degree in English. Shortly afterwards he started to visit the outer Cape and thus began his career as a self-taught architect. Hall not only purchased 180 acres on Bound Brook Island in 1937 and renovated, relocated, and recycled buildings on the Baker-Biddle land and other properties, he also created a welcoming atmosphere to the Modernist architects arriving in Wellfleet and Truro in the 1940s. These architects can be classified into two groups. One is the "Brahmin Bohemians" who were self-taught architects, well-educated from wealthy backgrounds. These included Hall, Jack Phillips, and Nathanial Saltonstall. The second group were the European

Modernists, some Bauhaus educated (McMahon and Cipriani 2014). Many of these Bauhaus architects had faced rejection of their art and persecution by the Nazis. In 1933 the Bauhaus school in Berlin was forced to close by the Nazis who suspected it of having ties to the Communist Party (Dyckhoff 2002). Eventually several Brahmin Bohemians and European Modernists settled in Wellfleet. This was a group of like-minded individuals who were eager to explore new ideas. Their summer homes served as their experimental laboratories. They were inexpensive; many were modular and so expandable. The architects used recycled and local material, open space interiors, and abundant windows for views and ventilation.

To a large degree it was Jack Hall of Bound Brook Island who created a welcoming atmosphere for these experimental architects - both the Brahmins and the Europeans. He also left a legacy of Modernist style buildings in Wellfleet one of which overlooks Cape Cod Bay on Bound Brook Island, the Hatch Cottage.

The Hatch Cottage was designed by Hall in 1960 for Robert Hatch, an editor of *The Nation* magazine, and his wife Ruth, an artist. The building sits on almost three acres with panoramic views over the dunes to Cape Cod Bay. It has modular rooms that are connected to each other by outdoor decks. Shutters can be opened or closed to regulate ventilation and light. The cottage respects the land around it and, as Peter McMahon of the Cape Cod Modern House Trust described, it seems as if the building could be picked up and moved without leaving a trace that it had existed in that place (McMahon and Cipriani 2014). This lightness and ephemeral quality was part of the philosophy of the Cape Cod Modernist Movement whose architects were influenced by simple vernacular buildings such as local oyster shacks. Homes were designed to be simple, inexpensive, experimental, and thoughtfully placed in the environment. The intimacy of these homes is an intrinsic part of their design. Today, the Hatch Cottage is used for events such as classes and tours and as a rental property. It is on the Massachusetts Historical Commission list of historic places as part of the Modernist House Movement.

During the mid-20th century Bound Brook Island, as well as some other remote areas of Wellfleet, became a haven for these intellectuals. They enjoyed Cape Cod Bay for recreation but the Herring River was no longer a focal point. It had long been severely restricted by the 1909 dike. For these new summer people, Bound Brook Island provided peace and quiet for contemplation, not a place whose resources offered them an economic base.

Chapter 8. Dike failure, controversial reconstruction, and ecological assessments, late 1960s through 2000

The leaking Chequesset Neck dike

By the late 1960s the 1909-built dike was clearly deteriorating, allowing more and more salt water to re-enter the flood plain upstream. Howard "Pokey" Snow, noted local oysterman and poet, discovered large numbers of oysters and soft-shell clams reestablishing as salinity and tidal exchange increased just above the dike. Cattails (*Typha angustifolia*) and other freshwater wetland vegetation above the dike were being killed by leaking saltwater and salt marsh grasses were recolonizing (Sterling 1976). However, the Chequesset Neck causeway was eroding, storm tides were reportedly causing flooding on the nearby Chequesset golf course, and the contentious debate over the dike was renewed.

Citing the ongoing restoration of shellfish populations above the failing dike, shellfishermen and conservationists, led by long-time resident, waterman, and writer Earle Rich (Fig. 24), petitioned the Town Meeting to overturn an earlier decision to rebuild the dike, and instead replace it with a bridge. The petition received 200 signatures and the support of the newly formed Wellfleet Conservation Commission, but the article failed at the June 1971 Town Meeting, which instead voted sixty-two to fifty-six to spend $37,500 as the town's share of the total $150,000 estimated to rebuild the dike.

In August of 1971 the Conservation Commission chair Stacy May and Shellfish Constable Bob Wallace sought the opinions of eminent Woods Hole (Massachusetts) salt marsh ecologists John Teal and Ivan Valiela, who supported tidal restoration and predicted little accompanying change in mosquitoes. Armed with this expert opinion, proponents of the bridge again petitioned for a Town Meeting vote for

Figure 24. Earle Rich, Wellfleet waterman, environmentalist, and author was for years a local leader in efforts to restore tidal exchange to the Herring River.

more study before the river was again diked. However, voters were told by town officials that the state might not cover the extra expense of a bridge, leaving the bill with the town. Special Town Meeting of 21 October 1971 voted 228 to 101 to rebuild the dike. Bridge proponents then petitioned the Massachusetts Department of Natural Resources (DNR; now Department of Environmental Protection) to study alternatives to a new dike, citing current wetland protection law that would never allow salt marsh diking. The state Division of Waterways postponed construction because of the opposition and asked DNR for an ecological study; on completion in September 1972 the DNR staff recommended a bridge (*Provincetown Advocate*, 7 September 1972).

Meanwhile the dike continued to leak, undermining the road, which was closed to traffic in 1972; residents of Griffin Island used High Toss and Duck Harbor Roads instead. At the same time, marine animals continued to re-establish above the dike, with one selectman reporting that "fishing had never been better."

Strengthened with new regulatory authority promulgated by the recently passed Wetlands Protection Act, the Wellfleet Conservation Commission held a hearing on the dike reconstruction project on 10 April 1973. Norton Nickerson, Tufts Biology professor and co-author of the Wetlands Protection Act, and Seashore Superintendent Leslie Arnberger argued for tidal restoration, though the latter cautioned more study. Oscar Doane, superintendent of the Cape Cod Mosquito Control Project, also supported tidal restoration, saying that mosquito breeding had decreased and the Project's work was made easier with more tidewater flowing through the failing dike. Later that month (26 April 1973), the Conservation Commission met again and approved rebuilding the dike but with two conditions; 1) tide heights must remain the same as presently, i.e., with the old dike failing, and 2) the herring run must be unimpeded by the new structure.

In support of ecosystem restoration, the Association for the Preservation of Cape Cod (APCC; now Association to Preserve Cape Cod), an environmental non-profit organization formed in 1968, shortly thereafter funded monitoring of existing conditions, including tide heights and salinity (Moody 1974) and vegetation (Snow 1975), to establish a base line prior to completion of the new structure. APCC released results of these studies in December 1974. Moody documented mean high- and low-tide heights and a tidal range of only 1.7 feet, indicating that water was being impounded upstream of the dike structure. Further evidence for freshwater impoundment, due to poor low-tide drainage, is Moody's salinity data showing that saltwater reached only 5000 feet upstream of the dike, well short of High Toss Road, despite the high river water levels relative to unrestricted Wellfleet Harbor. Snow reported that much of the cattail marsh immediately above the dike, established after the tides were blocked in 1909, was being recolonized by salt marsh plants, apparently due to the dike leakage and consequently increased salinity since about 1969.

Despite complaints from the new chairman of the Conservation Commission Kirk Wilkinson that the planned dike opening was too small to achieve the tide heights prescribed in the project's permit, Town Meeting voted in January 1974 for an additional $23,000 to cover escalating costs and reconstruction began in May of that year (Fig. 25). By the following January (1975) local oysterman Howard Snow reported that about 500 bushels of soft-shell clams and 18 bushels of oysters, which had established above the dike when it was leaking, died due to low salinity and low temperature (*The Cape Codder*, 9 January 1975). During the same winter, 18 bushels of oysters that had established above the leaking dike were removed and transplanted in the harbor. By March 1975 the new dike was completed, but shellfishermen testified in affidavits given to the Massachusetts Coastal Zone Management Office that increased sedimentation below the dike was covering oyster beds. This is

Figure 25. Reconstruction of the Chequesset Neck dike across the Herring River circa 1974. Rebuilding of the dike, and continued blockage of tidal exchange in Herring River, was even more controversial in the 1970s than when the original dike was proposed in 1906. This is expected given the growing appreciation for the values of coastal wetlands, along with local observations of shellfish re-establishment above the dike when the structure was leaking. Although the new tide valves were much more durable than those in the 1908 structure, the dike functioned the same in radically limiting seawater inflow during flood tides, and impeding drainage during ebb tides.

expected because the new dike severely reduced upstream flow of both water and fine sediment, the latter settling just seaward of the structure. Meanwhile, returning river herring reportedly accumulated on the harbor side of the new dike, unable to pass through to the river and their spawning ponds (Sterling 1976).

With the dike rebuilt, a follow-up study of vegetation above the dike by the state attorney general's office (Gaskell 1978) showed that the cattail and recovering salt marsh between the dike and High Toss Road observed by Snow (1975) had been mostly replaced by non-native common reed (*Phragmites australis*). This was likely due to the very small sluice gate opening maintained by the town during this period (1975-1978), keeping salinities too high for cattail, but low enough to favor *Phragmites* over native salt-marsh grasses. Common reed rapidly replaces freshwater wetland vegetation that has been killed by salt, so long as salinity remains below about 18 parts per thousand; cattail tolerates salinity up to only about 10 parts per thousand. (Note that the salinity of seawater in Cape Cod Bay is about 32 parts per thousand.) Interestingly, Snow's Figure 4 in her 1975 report shows only salt marsh grasses and cattail between the dike and High Toss Road and no *Phragmites*; that area now includes about 40 acres of dense and exclusive common reed.

Confusion and controversy reigned during the period 1975-1982 over the interpretation of APCC's tidal base-line data (Moody 1974) and the new dike opening needed to reproduce those tide heights per the Conservation Commission's conditions for dike reconstruction. When it was clear that tides were below prescribed levels, a group of concerned citizens sued the state Department of Public Works (DPW) to open the dike further; however, the town selectmen, who had received the dike crank and key after reconstruction, refused to open the sluice gate further and refused to relinquish the crank and padlock key to the state. Finally the state attorney general's office brought suit against the town for dike control to comply with the Conservation Commission's order of conditions under the Wetlands Protection Act. At the request of the state attorney general, Seashore staff systematically measured tide heights and found them well below prescribed levels (Portnoy 1981, 1982). Shortly thereafter the Town finally turned over the dike controls to the state Department of

Environmental Quality Engineering (now Department of Environmental Protection) which opened the 6-foot-wide sluice gate to a height of 24 inches, the height at which it remains today (2016).

In response to concerns that the new dike would damage the river herring run, the Massachusetts Division of Marine Fisheries (MA DMF) began a program in 1975 of stocking 1000-2000 "ripe" (spawning) alewives in Gull Pond each spring. This stocking continued until at least 1978 (letter from J.D. Fiske, MA DMF, to R.C. Gaskell, Department of Attorney General, 15 November 1978).

The dike and water quality

In February of 1980 Shellfish Constable Bob Wallace reported that the American eel (*Anguilla rostrata*) harvest in the river had dropped from several tons in 1976 and 1977 to 25 pounds in fall 1979 (The Advocate). In October of 1980, while conducting routine tide height and salinity surveys in Herring River, National Park Service Ecologists Michael Soukup and John Portnoy encountered thousands of large dead and dying eels from the upstream extent of seawater below High Toss Road to Route 6. Moribund eels that were moved from the diked river either to fresh pond water above Route 6 or into seawater below the dike recovered overnight, suggesting an environmental stress rather than disease. A subsequent investigation by the Seashore and the MA DMF showed that the fish kill was due to low pH, and the low pH was due to diking, drainage, and subsequent disturbance to natural chemical cycling, particularly involving sulfur (Soukup and Portnoy 1986). The pH in some drainage ditches was as low as lemon juice and low enough to leach aluminum from native clays in forms and concentrations toxic to fish.

In June of 1983 Portnoy reported to the Seashore Advisory Commission on the linkages between the diking and drainage in the

Herring River and surface-water acidification, fish kills, and high nuisance mosquito production; he recommended that the new dike be completely opened. At that meeting Superintendent Oscar Doane of Cape Cod Mosquito Control Project also supported increased tidal flow for mosquito control. Later during that summer, Seashore biologists first documented total dissolved oxygen depletion and massive juvenile herring mortality in the river main stem below Route 6 (Portnoy 1991). In 1986, after observing repeated summer oxygen depletions and fish kills, the Seashore began blocking herring emigration from the spawning ponds when monitoring showed insufficient dissolved oxygen in the river (Portnoy et al. 1987).

Meanwhile, the relief from biting mosquitoes promised by the dike proponents in 1906 had not materialized, at least during wet springs and summers. Based on field sampling begun in 1981, Seashore scientists concluded in 1984 that the most abundant human-nuisance mosquito species (*Ochlerotatus cantator*) bred abundantly in the diked and freshened Herring River flood plain (Portnoy 1984). Field scientists found dense concentrations of late-instar mosquito larvae and pupae (ready to hatch to flying adults) in the extremely acidic pools and ditches where estuarine fish, important predators of mosquito immatures, could not survive. It appeared that, instead of eliminating these pests, the radical solution of diking the river had in fact created extensive floodwater mosquito breeding habitat with few predators.

Although the Herring River estuary is within the Cape Cod National Seashore, in its agreement with the National Park Service, the Town of Wellfleet retained control of the shellfishing industry and management of shellfish resources within the estuary. In 1967, the shellfish beds in Town were mapped and by 1976, the shellfish department initiated propagation efforts in the River. For example, quahog seed was planted in 1979 and in the late 1980s, hundreds of bushels of oysters were transplanted from Chipman's Cove to Great Island. In 1985, the Department of Environmental Protection, then responsible for sampling the bacteriological quality of shellfish waters, classified the extensive oyster beds in Herring River mouth as "prohibited" to shellfish harvest due to fecal coliform contamination. It was later found that tidal restrictions in general concentrate coliform bacteria, probably due to reduced flushing and reduced salinity, the latter prolonging the survival of fecal coliform in the environment (Portnoy and Allen 2006). In 1987, with the efforts of the Natural Resource Task Force and the Massachusetts Department of Public Works, catch basins were installed in strategic areas in order to prevent stormwater runoff into the River with the goal of preventing conditional closures to shellfish harvest.

In 1991, after intensive lobbying efforts by both the Shellfish Constable and Health Agent, the MA DMF opened the Herring River to shellfishing for the first time in two years, on a rainfall closure basis from November 1 to December 15. The Herring warden reported that during this time, "commercial permit holders harvested 1498 bushels of oysters and 676 bushels of oyster seed." In 1992, precautionary closures after rainfall made the River off-limits for most of the fall. When the area was opened briefly, 550 bushels of quahogs and 2600 bushels of oysters were harvested, providing much needed relief for many commercial shellfishermen. However, the conditional closures began to last longer and longer and in 2005-2006, the MA DMF closed most of the River to shellfish harvest on a permanent basis. As of this writing (2016), the oyster beds immediately seaward of the dike remain closed to harvest; however, planned tidal restoration should greatly improve bacteriological water quality (Portnoy and Allen 2006).

The Town continued to manage the shellfish resources in the River by planting bay scallop shells to collect oyster spat, attempting to plant softshell clam seed, and moving seed oysters into the area. With the help of commercial harvesters, the Shellfish Department transplanted about 250 bushels of oysters from the closed area of the River to a relay site off Chequesset Neck Road, allowing the "contaminated" shellfish to purge themselves and be suitable for harvest after a few months. Efforts to transplant oysters from the closed area of the River to water suitable for harvest continued until 2008. This labor-intensive management technique has become controversial because of the possibility of spreading oyster diseases to other parts of the Harbor.

Scientific findings regarding the adverse effects of the dike on water quality (Soukup & Portnoy 1986, Portnoy 1991), fish, and wildlife habitat (Portnoy et al. 1987), and even, ironically because of the original rationale for diking, mosquito control (Portnoy 1984), prompted research into possible restoration alternatives and their likely environmental and social effects. By 1987, Rutgers University had completed an evaluation of hydrologic alternatives, based on hydrodynamic modeling, and established a base line inventory of vegetation, benthic invertebrates, fish, and birds (Roman 1987). Shortly thereafter, a study to address the potential for golf-course flooding with Herring River tidal restoration was conducted on Mill Creek, a tributary to Herring River (Nuttle 1990); the study recommended that, if tides were restored to Herring River, a dike should be built across Mill Creek to prevent tidal flooding. This was expected given that the Chequesset golf course was established on Mill Creek salt marshes about 1929 (twenty years after tides had been blocked by the dike at Chequesset Neck) with the addition of little if any fill to raise land-surface elevations.

By 1990, the Cape Cod Mosquito Control Project vocally supported tidal restoration to ease mosquito control, with the understanding that projected increases in tidal flushing and improvements in water quality would favor mosquito predators, especially fish. Also in 1990, the

Massachusetts Division of Fisheries and Wildlife formally urged tidal restoration to reestablish important wintering waterfowl habitat, especially for American black ducks (*Anas rubripes*), a high-priority species for conservation throughout the Atlantic flyway (Devers and Collins 2011).

The Herring River problem was somewhat quiescent through the 1990s, except for multiple public presentations of ecological and hydrological research, discussions with low-lying property owners, and continued research into the potential for tidal restoration. In 1991, the US Geological Survey, employing surface geophysical measurements and deep observation wells, determined that tidal restoration would not affect private water supply wells unless they were installed within the actual historic coastal flood plain (Fitterman and Dennehy 1991). Earlier work employing hydrodynamic modeling as a predictive basis for ecological restoration of both the Herring River and Hatches Harbor (Provincetown) salt marshes, was published in 1995 (Roman et al. 1995).

In the early 2000s, scientists from the University of Rhode Island and the Seashore conducted hydrodynamic, salinity and sediment-transport modeling to assess the likely effects of tidal restoration (Spaulding and Grilli 2001). For many years local shellfishermen and aquaculturists had expressed two concerns about reopening Herring River to tidal exchange: 1) the potential for fine sediment to be carried downstream by the newly restored tides, a process which might cover Wellfleet Harbor's extensive and lucrative shellfish beds and aquaculture grants; and 2) the potential for increased tidal flow to create a direct opening between the River and Cape Cod Bay through "The Gut" barrier beach connecting Great and Griffin Islands – this has been a long-held concern of some local shellfishermen because of the likelihood that such an opening would change the salinity and temperature regime of the harbor, and adversely affect shellfish beds. Spaulding and Grilli's work showed that, both under existing dike conditions and with full tidal restoration, flood tide velocities are faster than ebb tide velocities; therefore, net sediment transport must be upstream, not downstream onto shellfish beds. In addition, coastal geologist Amy Dougherty (2004) from Boston University used historic maps and aerial photographs to predict any changes in sedimentation; her results showed that sediment patterns and channel location even before the River was diked were similar to current conditions; therefore, she concluded that large changes in sediment transport would not be expected if tidal exchange were to be restored. Dougherty also addressed the likelihood of the restored River causing a breach for regular tidal exchange through The Gut. Her analysis of the area's geologic history, based on coastal survey maps, showed that the river occupied its current position even before the dike was built, and therefore flow from an unrestricted Herring River was too weak to maintain an opening through the Gut barrier beach. Dougherty presented her findings to both the Wellfleet Shellfish Advisory Board and Conservation Commission in the winter of 2004.

In 2003, based on many years of water chemistry research submitted by Seashore scientists and cooperators, the state Department of Environmental Protection listed the Herring River as "impaired", under the federal Clean Water Act, Section 303(d), for low pH, high metals, and fecal coliforms - all caused by the many decades of diking and drainage. It had become clear to most scientists, land managers and environmental regulators that the conditions within the diked Herring River, and many similarly altered coastal marshes throughout the world, were untenable for aquatic and marine life.

> **The "Gut"**
> The Gut is technically a tombolo, i.e, a sandbar connecting an island to the mainland or connecting two islands. The Gut, in the Herring River estuary, is a popular area for hiking, birding, swimming, fishing, and other forms of recreation. It is also a critical nesting habitat for endangered piping plovers (*Charadrius melodus*) and threatened diamondback terrapins. Because the Gut, connecting Griffin Island to Great Island is so narrow and exposed to west and northwest winds, it has been subject to erosion. This area has been of long concern to Wellfleet citizens, especially shellfishermen who fear that a breach in the Gut will affect shellfish resources in the Herring River. The Wellfleet Gut Committee was formed and met in 1977 to discuss methods to prevent erosion. The Gut Committee was active for several years, planting beach grass, and installing snow fencing. Because this area is part of the Cape Cod National Seashore and thus subject to the National Park Service (NPS) management plans, no further erosion prevention measures could be undertaken by the Gut Committee.

In 2004, hydrologists from the US Geological Survey (Masterson 2004) and the National Park Service (Martin 2004) completed numerical modeling assessments of the effects of restored tide heights and salinity in Herring River, as predicted by the Spaulding and Grilli hydrodynamic model, on the freshwater aquifer under adjacent developed uplands. The concern was for saltwater intrusion into domestic wells; however, the model simulations by Masterson and Martin showed no appreciable effect on the depth to the freshwater/saltwater interface in the areas adjacent to the River.

Chapter 9. Planning for tidal restoration, 2004 to the present

In the early 2000s, with growing scientific evidence, public presentations, and state, federal, and environmental organization support, consensus and interest in Wellfleet was building to at least re-examine the management of the diked Herring River. In August of 2005, the Wellfleet Board of Selectmen approved an agreement between the Town and the Seashore (Memorandum of Understanding I (MOU I)) to appoint both Herring River stakeholder and technical committees. The latter, representing the relevant boards and committees of the Towns of Wellfleet and Truro, the National Park Service, the US Fish and Wildlife Service, the Natural Resource Conservation Service, the National Marine Fisheries Service and the Massachusetts Wetland Restoration Program (now Division of Ecological Restoration), was charged with examining the technical merits of tidal restoration and developing a preliminary restoration plan. By 2006 news of the potential 1100-acre wetland restoration project in Wellfleet and Truro reached the offices of the Massachusetts congressional delegation; in October of 2006 Senator Edward Kennedy and Congressman William Delahunt addressed restoration partners at the Herring River dike to give their support and to describe their efforts to obtain funding.

The Herring River Technical Committee met from October 2005 through 2007 and produced a Conceptual Restoration Plan (CRP) (Herring River Technical Committee and ENSR 2007) and another memorandum of understanding (MOU II) for consideration by the towns and the Seashore. Based on the findings in the CRP, the new agreement acknowledged that tidal restoration was both feasible and in the public interest; it further established the Herring River Restoration Committee to replace the technical committee, albeit with the same agency representation, and to undertake all aspects of project planning.

The Herring River Restoration Committee has conducted monthly meetings from 2008 to the present (2016) and produced a Draft Environmental Impact Statement (EIS) in 2014, incorporating detailed restoration and mitigation plans. Fundamental to the EIS is hydrodynamic modeling: to describe how the diked system functions with respect to tide heights and salinity, and to predict changes to the estuary under a range of tidal restoration scenarios (Woods Hole Group 2012). Among the four alternatives considered, the Restoration Committee prefers almost full restoration within the Seashore boundary, but would limit tide heights in the more developed sub-basins outside the Seashore, including Mill Creek and upper Pole Dike Creek. After the final EIS the project will move into the next phases: permitting, fund raising, and construction to restore tidal flow.

An early observation by the Restoration Committee was the need for the establishment of a private non-profit entity to raise funds and provide communication, education, and outreach for this complicated project. In 2009, the Friends of Herring River was formed, initially under the wing of the Friends of Cape Cod National Seashore. Now an independent non-profit organization, Friends of Herring River regularly engages in Restoration Committee discussions, applies for government and private grants for the project, manages consulting contracts, publicly presents and writes about the planned restoration in newsletters, fact sheets and social media, collaborates on school science programs, and volunteers to collect monitoring data.

Since the spring of 2009, the Friends have collaborated with the Association to Preserve Cape Cod (APCC) and the Massachusetts Division of Marine Fisheries (MA DMF) to conduct an annual sample-census of the herring run. Volunteers count immigrating adult fish over 10-minute blocks from the beginning of the upstream run in late March until late May. Data are submitted to APCC and MA DMF to produce a total population estimate. The annual total has varied from

7000 to 60,000 fish, a far cry from the 200,000 harvested at the old fishery station before the dike in the late 1800s; however, the hope is that tidal restoration will reduce obstructions to fish passage, improve water quality, and restore estuarine nursery habitat for juvenile river herring, leading to consistently higher annual counts.

Meanwhile, as some level of tidal restoration has begun to appear more likely, the need for scientific assessment has increased, especially in view of the long period over which the river flood plain has been isolated from tidal seawater. One hundred years of diking and consequent effects on water levels and water quality has radically changed the community of plants and animals within the Herring River flood plain. Some of the animals that presently occupy the freshened marshes above the dike are both rare (state-listed by the Massachusetts Natural Heritage and Endangered Species Program) and sensitive to restored tidal flow and salinity in the Herring River. Work is currently (2016) under way to develop and implement study plans for monitoring, documenting, and assessing the response of the following animals, listed as rare, threatened, or endangered by the Massachusetts Natural Heritage and Endangered Species Program: least bittern (*Ixobrychus exilis*), American bittern (*Botaurus lentiginosus*), northern harrier (*Circus cyaneus*), eastern box turtle (*Terrapene carolina carolina*), diamondback terrapin (*Malaclemys terrapin*) and water willow stem-borer moth (*Papaipema sulphurata*).

Besides these site- and species-specific assessments, the estuary's long period of ecological disturbance and imminent restoration provides a rare opportunity for research in a wide range of currently important resource-management topics. The US Geological Survey (USGS) has several studies under way. Researchers from USGS and the University of Massachusetts are marking herring with passive integrated transponder (PIT) tags to see how the dike and upstream culverts might be impeding their passage to and from the spawning ponds (Castro-Santos 2013). USGS and Cape Cod National Seashore hydrologists and chemists are conducting continuous long-term water-quality monitoring at the dike structure to predict and follow changes in water chemistry, given the long period over which natural estuarine chemical cycling has been disturbed. Finally, perhaps the most timely and forward looking research initiative is being conducted by the USGS and the Marine Biological Laboratory (Woods Hole, MA) scientists assessing the effects of diking, drainage, and seawater restoration on sulfur, nitrogen, and carbon cycling, and especially carbon storage, in Herring River marshes (Colman et al. 2015). In this way, lessons learned through the 100-year "experiment" of diking, and hopefully restoring, the Herring River estuary will contribute significantly to our understanding of the functions and values of coastal wetlands throughout the world.

Chapter 10. Conclusion

Natural resources can not only be overused and depleted but also forgotten. In this "environmental amnesia," each successive human generation develops its own perspective and (usually lower) expectations of what is normal for the quality of the land, sea, and air that surround and support us. Unfortunately, that base line for normalcy in the case of our coastal resources has been a decidedly degrading one. We believe that the history of Wellfleet and its Herring River provides a particularly good example of this for coastal New England.

A first step in environmental restoration is understanding what natural resource values have been lost, and how it could have happened. For example, it is very useful to know what people were thinking when they made the decision to block tides from the Herring River. What was the stated goal; what were the social and economic forces involved in pushing the project; and what did people know of the likely ecological effects?

But of course the wisdom of the decision to cut off the Herring River estuary from the marine environment is too important to be judged by 19th century knowledge; as an analogy, few people today would opt for 19th century medicine for critical health decisions. Scientific research over the past century has identified major estuarine functions and social benefits that were very poorly understood 100 years ago when so many of New England's salt marshes were diked, drained, and filled. Importantly, modern ecological knowledge is cumulative - the antithesis of environmental amnesia. We hope that this book not only provides the general reader with insight into the reasoning and actions of those who have come before us, but also gives ecologists important environmental background as they restore the plants and animals of the tidal Herring River.

Herring River Restoration

Its mouth will widen over years.
Water will flow and overflow
in places it has not been
in over a century.
Will herring again run
in millions?
Will we fish once more
at Bound Brook Island
below the Atwood-Higgins house,
near the shimmering birches?

Sharon Dunn

Literature Cited

Army Corp of Engineers. 1908. Hearing 10 June 1908 on petition by Board of Harbor and Land Commissioners of Massachusetts for permission of Secretary of War to build a dike across Herring River, Wellfleet, Massachusetts, Lieutenant Colonel Edward Burr, ACOE, Massachusetts District, presiding.

Belding, D.L. 1920. Report on the alewife fishery of Massachusetts. Commonwealth of Massachusetts. Department of Conservation. Division of Fisheries and Game.

Biddle, F. 1962. In Brief Authority. Doubleday & Company, Garden City, NY.

Bradley, J.W., Survey Director. 1986. Historic and Archeological Resources of Cape Cod and The Islands: A Framework for Preservation Decisions. The Massachusetts Historical Commission. Boston, MA.

Braginton-Smith, J. and D. Oliver. 2008. Cape Cod Shore Whaling. The History Press. Charleston, SC.

Braun, E.K. and D.P. Braun. 1994. The First Peoples of the Northeast. Moccasin Hill Press. Boston, MA.

Burke, W.P. 2009. Atwood-Higgins House: The Debate over George Higgins's Whimsical Farmstead on Bound Brook Island. Cape Cod National Seashore. Park News. p.6.

Burling, F.P. 1978. The Birth of Cape Cod National Seashore. The Leyden Press. Plymouth, MA.

Cape Cod National Seashore and Herring River Restoration Committee. 2015. Herring River Restoration Project Final Environmental Impact Statement/Environmental Impact Report.

Castro-Santos, T. 2013. Migratory delay, predation, and passage success of anadromous river herring and other aquatic organisms in coastal rivers and estuaries of the Cape Cod National Seashore. Proposal to the National Park Service and US Geological Survey.

Colman, J., K. Kroeger, T. Smith and K. Lee. 2015. Water-quality, the controlling factor on the herring run, aquaculture, and blue carbon at the Herring River salt-marsh restoration, Cape Cod National Seashore. Proposal to the National Park Service and US Geological Survey.

Cumbler, J.T. 2014. Cape Cod: An Environmental History of a Fragile Ecosystem. University of Massachusetts Press. Amherst, MA.

Devers, P.K. and B. Collins. 2011. Conservation Action Plan for the American Black Duck, First Edition. US Fish & Wildlife Service Division of Migratory Bird Management.

Deyo, S.L. 1890. History of Barnstable, Massachusetts. H.W. Blake Co. New York, NY

Donaldson, E., L., H. Laham, and M. C. Brown. 2010. Atwood-Higgins Historic Cultural Landscape Report and Out Buildings Historic Structures Report: Cape Cod National Seashore, Wellfleet MA National Park Service.

Dougherty, A. J. 2004. Sedimentation concerns associated with the proposed restoration of Herring River marsh, Wellfleet, Massachusetts. Report to Cape Cod National Seashore, Town of Wellfleet and Association of Women Geoscientists.

Dyckhoff, T. 2002. Mies and the Nazis, The Guardian. November 30.

Echeverria, D. 1993. A History of Billingsgate. The Wellfleet Historical Society, Wellfleet, MA.

Fiske, J.D. 1978. Massachusetts Division of Marine Fisheries. Letter to R.C. Gaskell, Massachusetts Department of Attorney General. 15 November.

Fitterman, D.V. and K.F. Dennehy (USGS). 1991. Verification of geophysically determined depths to saltwater near the Herring River (Cape Cod National Seashore), Wellfleet, Massachusetts. 1991 June 4; Open-File Report 91-321.

Gaskell, R.C. 1978. Survey of existing biological conditions east of the Herring River dike, Wellfleet. Report to the Massachusetts Attorney General's Office.

Gillis, N.A. and H. Herbster. 2013. Endangered Archeological Sites Project Archaeological Reconnaissance and Intensive Survey – Great Island and Great Beach Hill, Cape Cod National Seashore, The Public Archaeology Laboratory, Inc. Pawtucket, RI.

Hall, J.H. 2013. Bound Brook Revealed. The Cape Cod Voice. February p. 13-26.

Herring River Technical Committee and ENSR. 2007. Herring River Tidal Restoration Project, Conceptual Restoration Plan. Prepared October 2007 for Towns of Wellfleet and Truro and Cape Cod National Seashore.

Historic American Buildings Survey/Historic American Engineering Record/Historic American Landscape Survey. (n.d.). Department of Interior National Park Service. Washington DC.

Holmes, R.D., C.D. Hertz, and M.T. Mulholland. 1995. Historic Cultural Land Use Study of Lower Cape Cod. The University of Massachusetts Archaelogical Services. Amherst, MA.

Hurd, S.H. (compiler). 1884. History of Plymouth County, Massachusetts.

Johnson, E.S. 1997. Archeological Overview and Assessment of the Cape Cod National Seashore, Massachusetts. University of Massachusetts, Amherst, MA.

Justice, B. 2013. The Founding Fathers, Education, and the "Great Contest." Palgrave McMillan. New York, NY.

Kino, C. 2014. Saving Modernism in Cape Cod. The Wall Street Journal. May 29.

Kittredge, H.C. 1930. Cape Cod: Its People and Their History. Houghton Mifflin Co. Boston, MA.

Lombardo, D. 2000. Wellfleet: A Cape Cod Village. Arcadia Publishing. Charleston, SC.

Martin, L. 2004. Salt marsh restoration at Herring River – An assessment of potential salt water intrusion in areas adjacent to Herring River and Mill Creek, Cape Cod National Seashore: Technical Report NPS/NRWRD/NRTR-2004/319.

Massachusetts Department of Public Works. 1923. Proposed Channel in Herring River, Wellfleet. Division of Waterways and Public Lands.

Massachusetts Historical Commission. 1984. Reconnaissance Survey Town Report: Wellfleet, MA.

Massachusetts Historical Commission Archaeological Exhibits Online
http://www.sec.state.ma.us/mhc/mhcarchexhibitsonline/intro.htm

Masterson, J.P. 2004. Simulated interaction between freshwater and saltwater and effects of ground-water pumping and sea-level change, Lower Cape Cod aquifer system, Massachusetts: US Geological Survey Scientific Investigations Report 2004-5014.

McKenzie, M. 2011. Clearing the Coastline: The Nineteenth-Century Ecological and Cultural Transformation of Cape Cod. University Press of New England. Hanover, NH.

McManamon, F. (ed.) 1984. Chapters in the Archeology of Cape Cod I: Results of the Cape Cod National Seashore Archeological Survey 1979-1981. Division of Cultural Resources North Atlantic Regional Resources North Atlantic Regional Office National Park Service. Boston, MA

McManamon, F. (ed.), 1985. Chapters in the Archeology of Cape Cod III: The Historic Period and Historic Period Archeology. Division of Cultural Resources North Atlantic Regional Office National Park Service. Boston, MA.

McMahon, P. and C. Cipriani. 2014. Cape Cod Modern: Midcentury Architecture on the Outer Cape. Metropolis Books. New York, NY.

Moody, J. 1974. Summary of Herring River estuary tidal and salinity data. Report to Association for the Preservation of Cape Cod, Orleans, MA.

Nuttle W.K. 1990. Extreme values of discharge for Mill Creek and options to control flooding from the Herring River. Report to National Park Service, Cape Cod National Seashore.

Nye, E.I. 1920. History of Wellfleet from Early Days to Present Time. F.B. and F.P. Goss. Hyannis, MA.

Oldale, R.N. 1992. Cape Cod and the Islands, the geologic story. Parnassus Imprints, East Orleans, MA.

Palmer, A.P. 1887. A Brief History of the Methodist Episcopal Church in Wellfleet Mass. Franklin Press, Rand, Avery and Co. Boston, MA.

Portnoy, J.W. 1981. Tidal water levels above the Herring River dike, Wellfleet, Massachusetts, 24 October through 16 November 1980. National Park Service.

Portnoy, J.W. 1982. Follow-up report: Tidal water levels above the Herring River dike, Wellfleet, Massachusetts, 28 December 1981 through 10 January 1982. National Park Service.

Portnoy, J.W. 1984. Salt marsh diking and nuisance mosquito production on Cape Cod, Massachusetts. Journal of the American Mosquito Control Association 44:560-564.

Portnoy, J.W. 1991. Summer oxygen depletion in a diked New England estuary. Estuaries 14(2):122-129.

Portnoy, J.W. and J.R. Allen. 2006. Effects of tidal restrictions and potential benefits of tidal restoration on fecal coliform and shellfish-water quality. Journal of Shellfish Research 25:609-617.

Portnoy, J.W., C. Phipps and B.A. Samora. 1987. Mitigating the effects of oxygen depletion on Cape Cod National Seashore anadromous fish. Park Science 8:12-13.

Portnoy, J.W. and M.A. Soukup. 1982. From salt marsh to forest: The outer Cape's wetlands. The Cape Naturalist 11:28-34.

Portnoy, J.W., C.T. Roman, and M.A. Soukup. 1987. Hydrologic and chemical impacts of diking and drainage of a small estuary (Cape Cod National Seashore): Effects on wildlife and fisheries. Pp. 253-265 *in* Whitman, W. R. and Meredith, W. H. "A Symposium on Waterfowl and Wetlands Management in the Coastal Zone of the Atlantic Flyway"; 16-19 Sept. 1986; Wilmington, DE.

Portnoy, J.W., M.G. Winkler, P.R. Sanford, and C.N. Farris. 2001. Kettle Pond Data Atlas: Paleoecology and Modern Water Chemistry. Cape Cod National Seashore.

Pratt, E. 1844. A comprehensive History Ecclesiastical and Civil of Eastham, Wellfleet and Orleans, County of Barnstable, Massachusetts from 1644 to 1844. W.S. Fisher and Co., Yarmouth, MA.

Price, R.N. 1912. Holston Methodism from its Origin to the Present Time. Vol II 1804-1824. Publishing House of the M.E. Church, South, South & Lamar. Richmond, VA.

Rich, E. ca. 1971. Is the Herring River Doomed. (newspaper clipping, Wellfleet Historical Society and Museum).

Rich, E. 1973. Cape Cod Echoes. Thompson's Printing, Orleans, MA.

Rich, E. 1978. More Cape Cod Echoes. Salt Meadow Publishers. Orleans, MA.

Rigolot, C. (trans). 2011. Saint John Perse Intime Par Katherine Biddle Journal 1940-1970. Gallimard. Paris, France.

Rockmore, M. 1979. Documentary Review of the Historical Archeology of the Cape Cod National Seashore. United States Department of the Interior, Division of Cultural Resources North Atlantic Regional Office, National Park Service, Lowell MA.

Roman C.T. 1987. An evaluation of alternatives for estuarine restoration management: the Herring River ecosystem (Cape Cod National Seashore). Technical Report, National Park Service Cooperative Research Unit, Rutgers University, New Brunswick, NJ.

Roman, C.T, R.W. Garvine and J.W. Portnoy. 1995. Hydrologic modeling as a predictive basis for ecological restoration of salt marshes. Environmental Management 19:559-566.

Snow, A. 1975. Recolonization of salt marsh species at the Herring River marsh, Wellfleet, Massachusetts. Report to Association for the Preservation of Cape Cod and Hampshire College.

Soukup, M.A. and J.W. Portnoy. 1986. Impacts from mosquito control-induced sulphur mobilization in a Cape Cod Estuary. Environmental Conservation 13(1):47-50.

Spaulding, M.L. and A. Grilli. 2001. Hydrodynamic and salinity modeling for estuarine habitat restoration at Herring River, Wellfleet, Massachusetts. Report to National Park Service.

Sterling, D. 1976. Our Cape Cod Salt Marshes. Informational Bulletin No. 6. Association for the Preservation of Cape Cod.

Stetson, J. 1963. Wellfleet: A Pictorial History. Wellfleet Historical Society.

Thoreau, H.D. 1865. Cape Cod. University Press, Welch, Bigelow and Co., Cambridge, MA

Torp, L., K.Franz. and M. Palus. 2013. Phase I (Site Identification) Archeological Survey Baker-Biddle Property Cape Cod National Seashore Town of Wellfleet, Barnstable County Massachusetts (Cape Cod National Seashore). The Ottery Group. Olney, MD.

True, F.W. 1887. The pound-net fisheries of the Atlantic states. Part XI. In Goode, G.B. The Fisheries and Fishery Industries in the United States. Government Printing Office. Washington DC.

Trust for Public Land. 2011. Historic Property Conserved for Cape Cod National Seashore, March 1.

Whalen, R.F. 2007. Everyday Life in Truro from the Indians to the Victorians. The History Press. Charleston, SC.

Whitman, L. 1794. A Topographical Description of Wellfleet in the County of Barnstable. Collection of the Massachusetts Historic Society. First Series. Vol. 3.

Whitman, H.T. and C. Howard. 1906. Report of Mssrs. Whitman and Howard Civil Engineers on Proposed Dike at Herring River, Wellfleet, Massachusetts.

Winkler, M.G. 1985. A 12,000-year history of vegetation and climate for Cape Cod, Massachusetts. Quarternary Research 23:301-312.

Winkler, M.G. and P.R. Sanford. 1994. Development of the Gull Pond chain of lakes and the Herring River basin, Cape Cod National Seashore. Final Report to National Park Service.

Woods Hole Group. 2012. Herring River hydrodynamic modeling for estuarine habitat restoration, Wellfleet, Massachusetts. Final Report. Prepared for Herring River Restoration Committee.

About the Authors

John W. Portnoy, Ph.D. holds degrees in biology, wildlife biology and marine ecology. He worked for the National Park Service for nearly thirty years as an ecologist at Cape Cod National Seashore, where his research focused on the effects of tide restrictions on salt marshes including the Herring River. Studies ranged from water chemistry to mosquito breeding ecology to wetland biogeochemical cycling, and were applied to the restoration of diked and drained salt marshes on outer Cape Cod. He lives in Wellfleet.

Alice M. Iacuessa worked thirty years as an international educator in Germany, Venezuela, and England. In England she taught for twenty-one years at The American School in London and served as Social Studies Department Chair. She is a member of the Board of Directors of the Friends of Herring River, the Wellfleet Marina Advisory Committee, and volunteers at the Cape Cod National Seashore. She lives in Wellfleet and London, England.

Barbara A. Brennessel, Ph.D. is Professor Emerita of Biology at Wheaton College in Norton, Massachusetts. Her recent research is focused on the Diamondback Terrapin (*Malaclemys terrapin*). She wrote "Diamonds in the Marsh: A Natural History of the Diamondback Terrapin." She also served on the Wellfleet's Shellfish Advisory Board, and wrote "Good Tidings: The History and Ecology of Shellfish Aquaculture in the Northeast." Inspired by her volunteer work for Friends of Herring River in Wellfleet, she wrote "The Alewives' Tale: The Life History and Ecology of River Herring in the Northeast," and a book for children, "The Adventures of Allie the Alewife." She lives in Wellfleet.